THE FIELD GUIDE TO HUMAN ERROR INVESTIGATIONS

The Field Guide to Human Error Investigations

SIDNEY DEKKER
Linköping Institute of Technology, Sweden

ASHGATE

Published by
Ashgate Publishing Limited
Gower House
Croft Road
Aldershot
Hampshire GU11 3HR
England

Ashgate Publishing Company
Suite 420
101 Cherry Street
Burlington, VT 05401-4405
USA

Ashgate website: http://www.ashgate.com

Paperback edition reprinted 2002, 2003

British Library Cataloguing in Publication Data
Decker, Sidney
The field guide to human error investigations
1. System failures (Engineering) 2. Human engineering
3. Industrial accidents
I. Title
363.1'1

Library of Congress Control Number: 2001095458

ISBN 0 7546 1917 6 (Hbk)
ISBN 0 7546 1924 9 (Pbk)

Printed and bound in Great Britain by MPG Books Ltd, Bodmin, Cornwall

Contents

Preface

You are faced with an incident or accident that has a significant human contribution in it. What do you do? How do you make sense out of other people's controversial and puzzling assessments and actions? You basically have two options, and your choice determines the focus, questions, answers and ultimately the success of your probe, as well as the potential for progress on safety:

- You can see human error as the cause of a mishap. In this case "human error", under whatever label—loss of situation awareness, procedural violation, regulatory shortfalls, managerial deficiencies—is the conclusion to your investigation.
- You can see human error as the symptom of deeper trouble. In this case, human error is the starting point for your investigation. You will probe how human error is systematically connected to features of people's tools, tasks and operational/organizational environment.

The first is called the old view of human error, while the second—itself already 50 years in the making—is the new view of human error.

Table 0.1: Two views on human error

The old view of human error	The new view of human error
Human error is a cause of accidents	Human error is a symptom of trouble deeper inside a system
To explain failure, you must seek failure.	To explain failure, do not try to find where people went wrong.
You must find people's: inaccurate assessments, wrong decisions, bad judgments.	Instead, find how people's assessments and actions made sense at the time, given the circumstances that surrounded them.

This Field Guide helps you reconstruct the human contribution to system failure according to the new view. In Part II, it presents a method for how to "reverse engineer" the evolving mindset of people who were caught up in a complex, unfolding situation. The Field Guide also wants to make you aware of the biases and difficulties in understanding past puzzling behavior—which is what Part I is about.

PART I OF THE FIELD GUIDE

The first six chapters of The Field Guide talk about the old view of human error—the problems it holds, the traps it represents, and the temptations that can make you fall into them. These chapters help you understand:

- The bad apple theory: why throwing out a few bad apples does not get rid of the underlying human error problem;
- Reactions to failure: why the surprising nature of failure makes you revert easily to the bad apple theory;
- That there is no such thing as a root or primary cause: accidents are the result of multiple factors—each necessary and only jointly sufficient;
- That large psychological labels may give you the illusion of understanding human error but that they hide more than they explain;
- Why human error cannot be explained by going into the mind alone. You have to understand the situation in which behavior took place;
- Why human factors data need to be left in the context from which they came: cherry picking and micro-matching robs data of its original meaning.

PART II OF THE FIELD GUIDE

The last seven chapters show you that human error is not necessarily something slippery or something hard to pin down. They show you how to concretely "reverse engineer" human error, like other components that need to be put back together in a mishap investigation. It shows how to rebuild systematic connections between human behavior and features of the tasks and tools that people worked with, and of the operational and organizational environment in which they carried out their work. The Field Guide will encourage you to build a picture of:

- how a process and other circumstances unfolded around people;
- how people's assessments and actions evolved in parallel with their changing situation;
- how features of people's tools and tasks and organizational and operational environment influenced their assessments and actions.

The premise is that if you really understand the evolving situation in which people's behavior took place, you will understand the behavior that took place inside of it. Here is what the last seven chapters talk about:

- Human error as a symptom of deeper trouble. Connecting people's behavior with the circumstances that surrounded them points you to the sources of trouble and helps explain behavior;
- How and where to get human factors data: from historical sources, interviews and debriefings, and process recordings;
- A method for the reconstruction of people's unfolding mindset—this is the central part around which the rest of The Field Guide revolves;
- Patterns of failure: Directs you to various patterns of failure in complex dynamic worlds, including the contributions from new technology, the drift into failure through unremarkable repetition of seemingly innocuous acts, failures to adapt and adaptations that fail, and coordination breakdowns;
- Writing meaningful human factors recommendations;
- Learning from failure as ultimate goal of an investigation: failures represent opportunities for learning—opportunities that can fall by the wayside for a variety of reasons;
- Rules for in the rubble: these are steps for how to understand human error, wrapping up the most important lessons from the book.

PART I

The Old View of Human Error:

Human error is a cause of accidents

To explain failure,
investigations must seek failure

They must find people's inaccurate assessments, wrong decisions and bad judgments

1. The Bad Apple Theory

There are basically two ways of looking at human error. The first view could be called "the bad apple theory". It maintains that:

- Complex systems would be fine, were it not for the erratic behavior of some unreliable people (bad apples) in it;
- Human errors cause accidents: humans are the dominant contributor to more than two thirds of them;
- Failures come as unpleasant surprises. They are unexpected and do not belong in the system. Failures are introduced to the system only through the inherent unreliability of people.

This chapter is about the first view, and the following five are about the problems and confusion that lie at its root.

Every now and again, nation-wide debates about the death penalty rage in the United States. Studies find a system fraught with vulnerabilities and error. Some states halt proceedings altogether; others scramble to invest more in countermeasures against executions of the innocent.

The debate is a window on people's beliefs about the sources of error. Says one protagonist: "The system of protecting the rights of accused is good. It's the people who are administering it who need improvement: The judges that make mistakes and don't permit evidence to be introduced. We also need improvement of the defense attorneys."[1] The system is basically safe, but it contains bad apples. Countermeasures against miscarriages of justice begin with them. Get rid of them, retrain them, discipline them.

But what is the practice of employing the least experienced, least skilled, least paid public defenders in many death penalty cases other than systemic? What are the rules for judges' permission of evidence other than systemic? What is the ambiguous nature of evidence other than inherent to a system that often relies on eyewitness accounts to make or break a case?

Each debate about error reveals two possibilities. Error is either the result of a bad apple, where disastrous outcomes could have been avoided if somebody had paid a bit more attention or made a little more effort. In this view, we wonder how we can cope with the unreliability of the human element in our systems.

Or errors are the inevitable by-product of people doing the best they can in systems that themselves contain multiple subtle vulnerabilities; systems where risks and safety threats are not always the same; systems whose conditions shift and change over time. These systems themselves are inherent contradictions between operational efficiency on the one hand and safety (for example: protecting the rights of the accused) on the other. In this view, errors are symptoms of trouble deeper inside a system. Like debates about human error, investigations into human error mishaps face the choice. The choice between the bad apple theory in one of its many versions, or what has become known as the new view of human error.

A Boeing 747 Jumbo Jet crashed upon attempting to take-off from a runway that was under construction and being converted into a taxiway. The weather at the time was terrible—a typhoon was about to hit the particular island: winds were high and visibility low. The runway under construction was close and parallel to the intended runway, and bore all the markings, lights and indications of a real runway. This while it had been used as a taxiway for quite a while and was going to be officially converted at midnight the next day—ironically only hours after the accident. Pilots had complained about potential confusion for years, saying that by not indicating that the runway was not really a runway, the airport authorities were "setting a trap for a dark and stormy night". The chief of the country's aviation administration, however, claimed that "runways, signs and lights were up to international requirements" and that "it was clear that human error had led to the disaster." Human error, in other words, was simply the cause, and that was that. There was no deeper trouble of which the error was a symptom.

The ultimate goal of an investigation is to learn from failure. The road towards learning—the road taken by most investigations—is paved with intentions to follow the new view. Investigators intend to find the systemic vulnerabilities behind individual errors. They want to address the error-producing conditions that, if left in place, will repeat the same basic pattern of failure.

In practice, however, investigations often return disguised versions of the bad apple theory—in both findings and recommendations. They sort through the rubble of a mishap to:

- Find evidence for erratic, wrong or inappropriate behavior;
- Bring to light people's bad decisions; inaccurate assessments; deviations from written guidance;
- Single out particularly ill-performing practitioners.

Investigations often end up concluding how front-line operators failed to notice certain data, or did not adhere to procedures that appeared relevant after the fact. They recommend the demotion or retraining of particular individuals; the tightening of procedures or oversight. The reasons for regression into the bad apple theory are many. For example:

- Resource constraints on investigations. Findings may need to be produced in a few months time, and money is limited;
- Reactions to failure, which make it difficult not to be judgmental about seemingly bad performance;
- The hindsight bias, which confuses our reality with the one that surrounded the people we investigate;
- Political distaste of deeper probing into sources of failure, which may de facto limit access to certain data or discourage certain kinds of recommendations;
- Limited human factors knowledge on part of investigators. While wanting to probe the deeper sources behind human errors, investigators may not really know where or how to look.

In one way or another, The Field Guide will try to deal with these reasons. It will then present an approach for how to do a human error investigation—something for which there is no clear guidance today.

UNRELIABLE PEOPLE IN BASICALLY SAFE SYSTEMS

This chapter discusses the bad apple theory of human error. In this view on human error, progress on safety is driven by one unifying idea:

COMPLEX SYSTEMS ARE BASICALLY SAFE

THEY NEED TO BE PROTECTED FROM UNRELIABLE PEOPLE

Charges are brought against the pilots who flew a VIP jet with a malfunction in its pitch control system (which makes the plane go up or down). Severe oscillations during descent killed seven of their unstrapped passengers in the back. Significant in the sequence of events was that the pilots "ignored" the relevant alert light in the cockpit as a false alarm, and that they had not switched on the fasten seatbelt sign from the top of descent, as recommended by jet's procedures. The pilot oversights were captured on video, shot by one of the passengers who died not much later. The pilots, wearing seatbelts, survived the upset.[2]

To protect safe systems from the vagaries of human behavior, recommendations typically propose to:

- Tighten procedures and close regulatory gaps. This reduces the bandwidth in which people operate. It leaves less room for error.
- Introduce more technology to monitor or replace human work. If machines do the work, then humans can no longer make errors doing it. And if machines monitor human work, they can snuff out any erratic human behavior.
- Make sure that defective practitioners (the bad apples) do not contribute to system breakdown again. Put them on "administrative leave"; demote them to a lower status; educate or pressure them to behave better next time; instill some fear in them and their peers by taking them to court or reprimanding them.

In this view of human error, investigations can safely conclude with the label "human error"—by whatever name (for example: ignoring a warning light, violating a procedure). Such a conclusion and its implications supposedly get to the causes of system failure.

AN ILLUSION OF PROGRESS ON SAFETY

The shortcomings of the bad apple theory are severe and deep. Progress on safety based on this view is often a short-lived illusion. For example, focusing on individual failures does not take away the underlying problem. Removing "defective" practitioners (throwing out the bad apples) fails to remove the potential for the errors they made.

As it turns out, the VIP jet aircraft had been flying for a long time with a malfunctioning pitch feel system ('Oh that light? Yeah, that's been on for four months now'). These pilots inherited a systemic problem from the airline that operated the VIP jet, and from the organization charged with its maintenance.

In other words, trying to change your people by setting examples, or changing the make-up of your operational workforce by removing bad apples, has litte long-term effect if the basic conditions that people work under are left unamended.

Adding more procedures

Adding or enforcing existing procedures does not guarantee compliance. A typical reaction to failure is procedural overspecification—patching observed holes in an operation with increasingly detailed or tightly targeted rules, that respond specifically to just the latest incident. Is this a good investment in safety? It may seem like it, but by inserting more, more detailed, or more conditioned rules, procedural overspecification is likely to widen the gap between procedures and practice, rather than narrow it. Rules will increasingly grow at odds with the context-dependent and changing nature of practice.

The reality is that mismatches between written guidance and operational practice always exist. Think about the work-to-rule strike, a form of industrial action historically employed by air traffic controllers, or customs officials, or other professions deeply embedded in rules and regulations. What does it mean? It means that if people don't want to or cannot go on strike, they say to one another: "Let's follow all the rules for a change!" Systems come to a grinding halt. Gridlock is the result. Follow the letter of the law, and the work will not get done. It is as good as, or better than, going on strike.

Seatbelt sign on from top of descent in a VIP jet? The layout of furniture in these machines and the way in which their passengers are pressured to make good use of their time by meeting, planning, working, discussing, does everything to discourage people from strapping in any earlier than strictly necessary. Pilots can blink the light all they want, you could understand that over time it may become pointless to switch it on from 41,000 feet on down.

And who typically employs the pilot of a VIP jet? The person in the back. So guess who can tell whom what to do. And why have the light on only from the top of descent? This is hypocritical—only in the VIP jet upset discussed here was that relevant because loss of control occurred during descent. But other incidents with in-flight deaths have occurred during cruise. Procedures are insensitive to this kind of natural variability.

New procedures can also get buried in masses of regulatory paperwork. Mismatches between procedures and practice grow not necessarily because of people's conscious non-adherence but because of the amount and increasingly tight constraints of procedures.

The vice president of a large airline commented recently how he had seen various of his senior colleagues retire over the past few years. Almost all had told him how they had gotten tired of updating their aircraft operating manuals with new procedures that came out—one after the other—often for no other reason than to close just the next gap that had been revealed in the latest little incident. Faced with a growing pile of paper in their mailboxes, they had just not bothered. Yet these captains all retired alive and probably flew very safely during their last few years.

Adding a bit more technology

More technology does not remove the potential for human error, but relocates or changes it.

A warning light does not solve a human error problem, it creates new ones. What is this light for? How do we respond to it? What do we do to make it go away? It lit up yesterday and meant nothing. Why listen to it today?

What is a warning light, really? It is a threshold crossing device: it starts blinking when some electronic or electromechanical threshold is exceeded. If particular values stay below the threshold, the light is out.

If they go above, the light comes on. But what is its significance? After all, the aircraft has been flying well and behaving normally, even with the light on.

WHY IS THE BAD APPLE THEORY POPULAR?

Cheap and easy

So why would anyone adhere to the bad apple theory of human error? There are many reasons. One is that it is a relatively straightforward approach to dealing with safety. It is simple to understand and simple, and relatively cheap, to implement. The bad apple theory suggests that failure is an aberration, a temporary hiccup in an otherwise smoothly performing, safe operation. Nothing more fundamental, or more expensive, needs to be changed.

A patient died in an Argentine hospital because of an experimental US drug, administered to him and many fellow patients. The event was part of a clinical trial of a yet unapproved medicine eventually destined for the North American market. To many, the case was only the latest emblem of a disparity where Western nations use poorer, less scrupulous, relatively underinformed and healthcare-deprived medical testing grounds in the Second and Third World. But the drug manufacturer was quick to point out that "the case was an aberration" and emphasized how the "supervisory and quality assurance systems all worked effectively". The system, in other words, was safe—it simply needed to be cleansed of its bad apples. The hospital fired the doctors involved and prosecuters were sent after them with murder charges.[3]

Saving face

In the aftermath of failure, pressure can exist to save public image. Taking out defective practitioners is always a good start to saving face. It tells people that the mishap is not a systemic problem, but just a local glitch in an otherwise smooth operation.

Two hard disks with classified information went missing from the Los Alamos nuclear laboratory, only to reappear under suspicious circumstances behind a photocopier a few months later. Under pressure to assure that the facility was secure and such lapses extremely uncommon, the Energy Secretary attributed the incident to "human error, a mistake". The hard drives were probably misplaced out of negligence or inattention to security procedures, officials said. The Deputy Energy Secretary added that "the vast majority are doing their jobs well at the facility, but it probably harbored "a few bad apples" who had compromised security out of negligence.[4]

Personal responsibility and the illusion of omnipotence

Another reason to adhere to the bad apple theory of human error is that practitioners in safety-critical domains typically assume great personal responsibility for the outcomes of their actions. Practitioners get trained and paid to carry this responsibility, and are proud of it.

But the other side of taking this responsibility is the assumption that they have the authority, the power, to match it; to live up to it. The assumption is that people can simply choose between making errors and not making them—independent of the world around them. This, however, is an illusion of omnipotence. It is commonly entertained by children in their pre-teens, and by the airline captain who said, "If I didn't do it, it didn't happen."

Investigators are often practitioners themselves or have been practitioners, which can make it easy to overestimate the freedom of choice allotted to fellow practitioners.

The pilot of an airliner accepted a different runway with a more direct approach to the airport. The crew got in a hurry and made a messy landing that resulted in some minor damage to the aircraft. Asked why they accepted the runway, the crew cited a late arrival time and many connecting passengers on board. The investigator's reply was that real pilots are of course immune to those kinds of pressures.

The reality is that people are not immune to those pressures, and the organizations that employ them would not want them to be. People do not operate in a vacuum, where they can decide and act all-powerfully. To err or not to err is not a choice. Instead, people's work is subject to and constrained by multiple factors. Individual responsibility is not always matched by individual authority. Authority can be restricted by other people or parts in the system; by other pressures; other deficiencies.

In the VIP jet's case, it was found that there was no checklist that told pilots what to do in case of a warning light related to the pitch system. The procedure to avoid the oscillations would have been to reduce airspeed to less than 260 knots indicated. But the procedure was not in any manual. It was not available in the cockpit. And it's hardly the kind of thing you can think up on the fly.

WHAT IS NOT RIGHT WITH THIS STORY?

If you are trying to explain human error, you can safely bet that you are not there yet if you have to count on individual people's negligence or complacency; if your explanation still depends on a large measure of people not motivated to try hard enough.

Something was not right with the story of the VIP jet from the start. How, really, could pilots "ignore" a light for which there was no procedure available? You cannot ignore a procedure that does not exist. Factors from the outside seriously constrained what the pilots could have possibly done. Problems existed with this particular aircraft. No procedure was available to deal with the warning light.

Whatever label is in fashion (complacency, negligence, ignorance), if a human error story is complete only by relying on "bad apples" who lack the motivation to perform better, it is probably missing the real story behind failure, or at least large parts of it.

Local rationality

The point is, people in safety-critical jobs are generally motivated to stay alive, to keep their passengers, their patients, their customers alive. They do not go out of their way to deliver overdoses; to fly into mountainsides or windshear; to amputate wrong limbs; to convict someone innocent. In the end, what they are doing makes sense to them at that time. This, in human factors, is called the local rationality principle. People are doing reasonable things given their point of view and focus of attention; their knowledge of the situation; their objectives and the objectives of the larger organization they work for. People do want to be bothered: their lives and livelihoods are on the line.

But safety is never the only concern, or even the primary concern. Systems do not exist to be safe, they exist to make money; to render a service; provide a product. Besides safety there are multiple other objectives—pressures to produce; to not cost an organization unnecessary money; to be on time; to get results; to keep customers happy. People's sensitivity to these objectives, and their ability to juggle them in parallel with demands for safety, is one reason why they were chosen for the jobs, and why they are allowed to keep them.

In the Los Alamos nuclear research facility, complacency was no longer a feature of a few individuals—if it ever had been. Under pressure to perform daily work in a highly cumbersome context of checking, double-checking and registering the use of sensitive materials, "complacency" (if one could still call it that) had become a feature of the entire laboratory. Scientists routinely moved classified material without witnesses or signing logs. Doing so was not a sign of malice. The practice had grown over time, bending to production pressures from which the laboratory owed its existence.[5]

When conducting a human error investigation, you have to assume that people were doing reasonable things given their circumstances. People were doing their best given the complexities, dilemmas, trade-offs and uncertainty that surrounded them. Just finding and highlighting people's mistakes explains nothing. Saying what people did not do does not explain why they did what they did.

The point of a human error investigation is to understand why actions and assessments that are now controversial, made sense to people at the time. You have to push on people's mistakes until they make sense—relentlessly. You have to reconstruct, rebuild their circumstances; resituate the controversial actions and assessments in the flow of behavior of which they were part and see how they reasonably fit the world that existed around people at the time. For this, you can use all the tools and methods that the Field Guide provides.

Notes

1 International Herald Tribune, 13 June 2000.
2 FLIGHT International, 6-12 June 2000.
3 International Herald Tribune, 22 December 2000 (p. 21).
4 International Herald Tribune, 19 June 2000.
5 International Herald Tribune, 20 June 2000.

2. Reacting to Failure

Have you ever caught yourself asking, "How could they not have noticed?", or, "How could they not have known?"? Then you were reacting to failure.

TO UNDERSTAND FAILURE, WE MUST FIRST UNDERSTAND OUR REACTIONS TO FAILURE

We all react to failure. In fact, our reactions to failure often make that we see human error as the cause of a mishap; they promote the bad apple theory. Failure, or people doing things with the potential for failure, is generally not something we expect to see. It surprises us; it does not fit our assumptions about the system we use or the organization we work in. It goes against our beliefs and views. As a result, we try to reduce that surprise—we react to failure.

A Navy submarine crashed into a Japanese fishing vessel near Hawaii, sinking it and killing nine Japanese men and boys. The submarine, on a tour to show civilians its capabilities, was demonstrating an "emergency blow"—a rapid re-surfacing. Time had been running short and the crew, crowded in the submarine's control room with sixteen visitors, conducted a hurried periscope check to scan the ocean surface. Critical sonar equipment onboard the submarine was inoperative at the time.

The commander's superior, an Admiral, expressed shock over the accident. He was puzzled, since the waters off Hawaii are among the easiest areas in the world to navigate. According to the admiral, the commander should not have felt any pressure to return on schedule. At one of the hearings after the accident, the Admiral looked at the commander in the courtroom and said "I'd like to go over there and punch him for not taking more time". The commander alone was to blame for the accident—civilians onboard had nothing to do with it, and neither had inoperative sonar equipment. [1]

Reactions to failure, such as in the example above, share the following features:

- **Retrospective**. Reactions arise from our ability to look back on a sequence of events, of which we know the outcome;
- **Proximal**. They focus on those people who were closest in time and space to causing or potentially preventing the mishap;
- **Counterfactual**. They lay out in detail what these people could have done to prevent the mishap;
- **Judgmental**. They say what people should have done, or failed to do, to prevent the mishap.

Reactions to failure interfere with our understanding of failure. The more we react, the less we understand. But when you look closely at findings and conclusions about human error, you can see that they are often driven by reactions to failure, and written in their language.

RETROSPECTIVE

Looking back on a sequence of events, knowing the outcome

> **INVESTIGATIONS AIM TO EXPLAIN A PART OF THE PAST**
>
> **YET ARE CONDUCTED IN THE PRESENT, AND THUS INEVITABLY INFLUENCED BY IT**

As investigator, you are likely to know:

- The outcome of a sequence of events you are investigating;
- Which cues and indications were critical in the light of the outcome—what were the signs of danger?
- Which assessments and actions would have prevented the outcome.

A highly automated airliner crashed on a golf course short of the runway at an airport in India. During the final approach, the aircraft's automation had been in "open descent mode", which manages airspeed by pitching the nose up or down, rather than through engine power. When they ended up too low on the approach, the crew could not recover in time. In hindsight, the manufacturer of the aircraft commented that "the crew should have known they were in open descent mode". Once outside observers learned its importance, the question became how the crew could have missed or miss-understood such a critical piece of information.

One of the safest bets you can make as an investigator or outside observer is that you know more about the incident or accident than the people who were caught up in it—thanks to hindsight:

- Hindsight means being able to look back, from the outside, on a sequence of events that led to an outcome you already know about;
- Hindsight gives you almost unlimited access to the true nature of the situation that surrounded people at the time (where they actually were versus where they thought they were; what state their system was in versus what they thought it was in);
- Hindsight allows you to pinpoint what people missed and shouldn't have missed; what they didn't do but should have done.

Hindsight biases your investigation towards items that you *now* know were important ("open descent mode"). As a result, you may assess people's decisions and actions mainly in the light of their failure to pick up this critical piece of data. It artificially narrows your examination of the evidence and potentially misses alternative or wider explanations of people's behavior.

Inside the tunnel

Look at figure 2.1. You see an unfolding sequence of events there. It has the shape of a tunnel which is meandering its way to an outcome. The figure shows two different perspectives on the pathway to failure:

- **The perspective from the outside and hindsight** (typically your perspective). From here you can oversee the entire sequence of events—the triggering conditions, its various twists and turns, the outcome, and the true nature of circumstances surrounding the route to trouble.

- **The perspective from the inside of the tunnel**. This is the point of view of people in the unfolding situation. To them, the outcome was not known (or they would have done something else). They contributed to the direction of the sequence of events on the basis of what they saw on the *inside* of the unfolding situation. To understand human error, you need to attain this perspective.

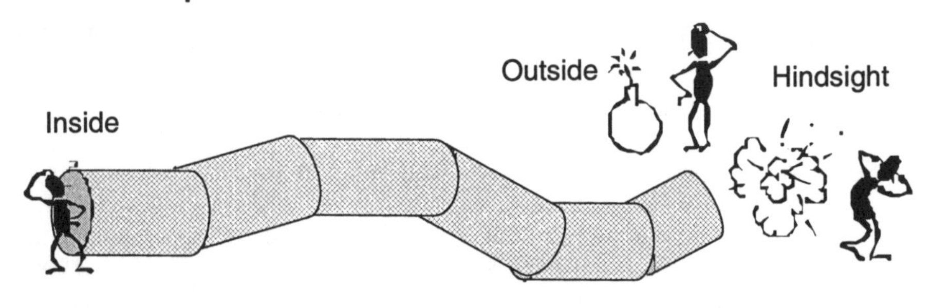

Figure 2.1: Different perspectives on a sequence of events: Looking from the outside and hindsight you have knowledge of the outcome and dangers involved. From the inside, you may have neither.

The Field Guide invites you to go inside the tunnel of figure 2.1. It will help you understand an evolving situation from the point of view of the people inside of it, and see why their assessments and actions made sense at the time.

Hindsight is everywhere

Hindsight is baked deeply into the language of accident stories we tell one another. Take a common problem today—people losing track of what mode their automated systems are operating in. This happens in cockpits, operating rooms, process control plants and many other workplaces. In hindsight, when you know how things developed and turned out, this problem is often called "losing mode awareness". Or, more broadly, "loss of situation awareness". What are we really saying? Look at figure 2.2. Loss of situation awareness is the difference between:

- what you *now* know the situation actually was like;
- what people understood it to be at the time.

It is easy to show that people at another time and place did not know what you know today ("they should have known they were in open descent mode"). But it is not an explanation of their behavior.

You must guard yourself against mixing your reality with the reality of the people you are investigating. Those people did not know there was going to be a negative outcome, or they would have done something else. It is impossible for people to assess their decisions or incoming data in light of an outcome they do not yet know about.

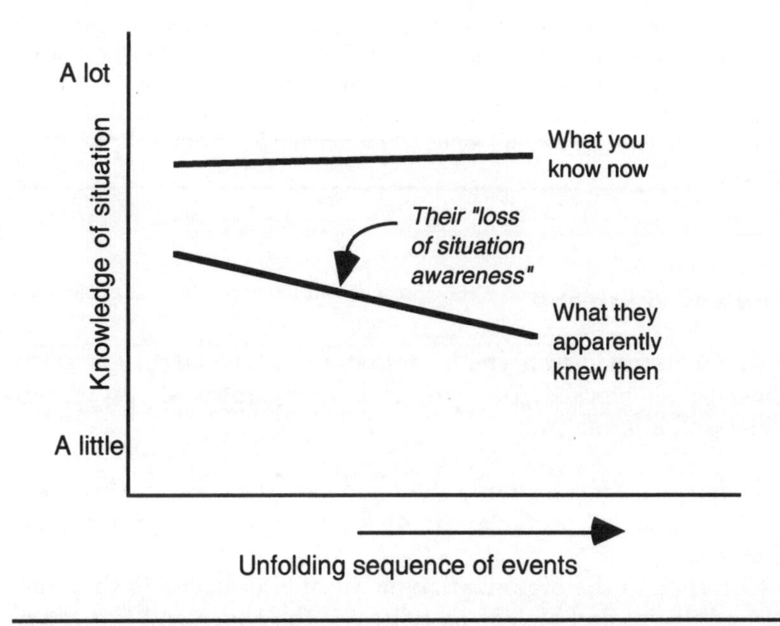

Figure 2.2: Hindsight is everywhere. Here, "loss of situation awareness" as the difference between your knowledge today of which aspects in the situation were critical, and what people apparently knew then.

PROXIMAL

Focusing on people at the sharp end

Reactions to failure focus firstly and predominantly on those people who were closest to producing and to potentially avoiding the mishap. It is easy to see these people as the engine of action. If it were not for them, the trouble would not have occurred.

Someone called me on the phone, demanding to know how it was possible that train drivers ran red lights. Britain had just suffered one of its worst rail disasters—this time at Ladbroke Grove near Paddington station in London. A commuter train had run head-on into a high-speed intercity coming from the other direction. Many travelers were killed in the crash and ensuing fire. The investigation returned a verdict of "human error". The driver of the commuter train had gone right underneath signal 109 just outside the station, and signal 109 had been red, or "unsafe". How could he have missed it? A photograph published around the same time showed sensationally how another driver was reading a newspaper while driving his train.

Blunt end and sharp end

In order to understand error, you have to examine the larger system in which these people worked. You can divide an operational system into a sharp end and a blunt end:

- At the sharp end (for example the train cab, the cockpit, the surgical operating table), people are in direct contact with the safety-critical process;
- The blunt end is the organization or set of organizations that supports and drives and shapes activities at the sharp end (for example the airline or hospital; equipment vendors and regulators).

The blunt end gives the sharp end resources (for example equipment, training, colleagues) to accomplish what it needs to accomplish. But at the same time it puts on constraints and pressures ("don't be late, don't cost us any unnecessary money, keep the customers happy"). Thus the blunt end shapes, creates, and can even encourage opportunities for errors at the sharp end. Figure 2.3 shows this flow of causes through a system. From blunt to sharp end; from upstream to downstream; from distal to proximal. It also shows where the focus of our reactions to failure is trained: on the proximal.

Proximal:

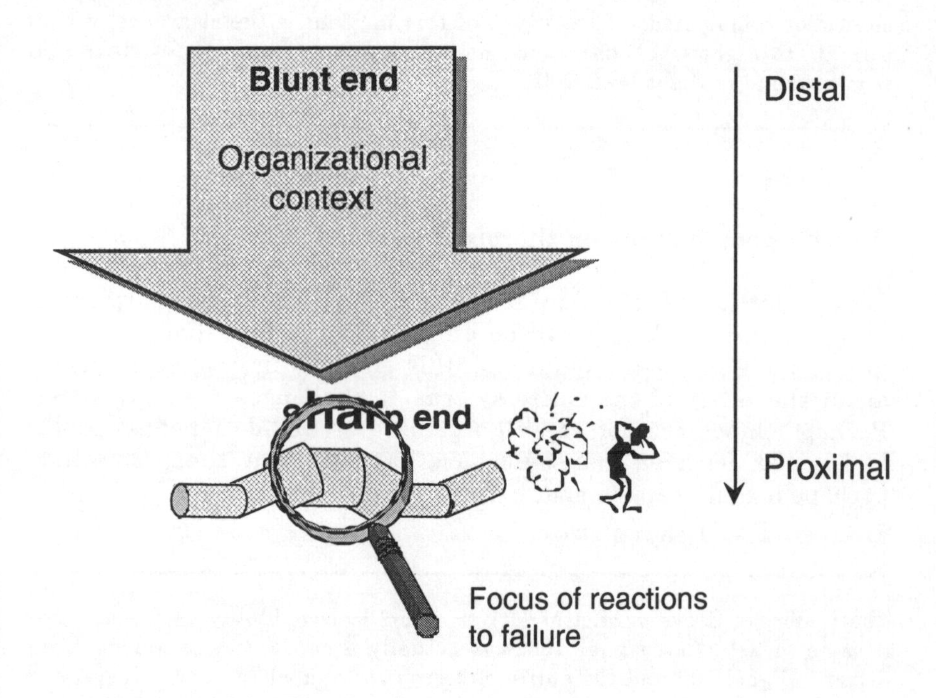

Figure 2.3: Failures can only be understood by looking at the whole system in which they took place. But in our reactions to failure, we often focus on the sharp end, where people were closest to causing or potentially preventing the mishap.

I was participating in an air traffic control investigation meeting that looked into a loss of separation between two aircraft (they came too close together in flight). When the meeting got to the "probable cause", some controllers proposed that the cause be "the clearance of one of these aircraft to flight level 200"—after all, that was the cardinal mistake that "caused" the separation loss. They focused, in other words, on the last proximal act that *could* have avoided the incident, but didn't, and thus they labeled it "the cause".

Such focus on the proximal is silly, of course: many things went before and into that particular clearance, which itself was only one act out of a whole stream of assessments and actions, set in what turned out to be a challenging and non-routine situation. And who says it was the last act that could have avoided the incident? What if the aircraft could have seen each other visually, but didn't? In that case there would be another proximal cause: the failure of the aircraft to visually identify one another. One controller commented: "if the cause of this incident is the clearance of that aircraft to flight level 200, then the solution is to never again clear that whole airline to flight level 200".

Why do people focus on the proximal?

Looking for sources of failure far away from people at the sharp end is counterintuitive. And it can be difficult. If you find that sources of failure lie really at the blunt end, this may call into question beliefs about the safety of the entire system. It challenges previous views. Perhaps things are not as well-organized or well-designed as people had hoped. Perhaps this could have happened any time. Or worse, perhaps it could happen again.

The Ladbroke Grove verdict of "driver error" lost credibility very soon after it came to light that signal 109 was actually a cause célèbre among train drivers. Signal 109 and the entire cluttered rack on which it was suspended together with many other signals, were infamous. Many drivers had passed an unsafe signal 109 over the preceding years and the drivers' union had been complaining about its lack of visibility.

In trains like the one that crashed at Ladbroke Grove, automatic train braking systems (ATB) had not been installed because they had been considered too expensive. Train operators had grudgingly agreed to install a "lite" version of ATB, which in some sense relied as much on driver vigilance as the red light itself did.

Reducing surprise by pinning failure on local miscréants

Some people and organizations see surprise as an opportunity to learn. Failures offer them a window through which they can see the true internal workings of the system that produced the incident or accident. These people and organizations are willing to change their views, to modify their beliefs about the safety or robustness of their system on the basis of what the system has just gone through. This is where real learning about failure occurs, and where it can create lasting changes for the good. But such learning does not come easy. And it does not come often. Challenges to existing views are generally uncomfortable. Indeed, for most people and organizations, coming face to face with a mismatch between what they believed and what they have just experienced is difficult. These people and organizations will do anything to reduce the nature of the surprise.

Some fighter pilots are not always kind on the reputation of a comrade who has just been killed in an accident. Sociologists have observed how his or her fellow pilots go to the bar and drink to the fallen comrade's misfortune, or more likely his or her screw-up, and put the drinks on his or her bar tab. This practice is aimed at highlighting or inventing evidence for why s/he wasn't such a good pilot after all. The transformation from "one of themselves" into "a bad pilot" psychologically shields those who do the same work from equal vulnerability to failure.

People and organizations often want the surprise in the failure to go away, and with it the challenge to their views and beliefs. The easiest way to do this is to see the failure as something local, as something that is merely the problem of a few individuals who behaved in uncharacteristic, erratic or unrepresentative (indeed, locally "surprising") ways.

Potential revelations about systemic vulnerabilities were deflected by pinning failure on one individual in the case of Oscar November.[2] Oscar November was one of the airline's older Boeing 747 "Jumbojets". It had suffered earlier trouble with its autopilot, but on this morning everything else conspired against the pilots too. There had been more headwind than forecast, the weather at the destination was very bad, demanding an approach for which the co-pilot was not qualified but granted a waiver, while he and the flight engineer were actually afflicted by gastrointestinal infection. Air traffic control turned the big aircraft onto a tight final approach, which never gave the old autopilot enough time to settle down on the right path. The aircraft narrowly missed a building near the airport, which was shrouded in thick fog. On the next approach it landed without incident.

Oscar November's captain was taken to court to stand trial on criminal charges of "endangering his passengers" (something pilots do every time they fly, one fellow pilot quipped). The case centered around the crew's "bad" decisions. Why hadn't they diverted to pick up more fuel? Why hadn't they thrown away that approach earlier? Why hadn't they gone to another arrival airport? These questions trivialized or hid the organizational and operational dilemmas that confront crews all the time. The focus on customer service and image; the waiving of qualifications for approaches; putting more work on qualified crewmembers; heavy traffic around the arrival airport and subsequent tight turns; trade-offs between diversions in other countries or continuing with enough but just enough fuel. And so forth.

The vilified captain was demoted to co-pilot status and ordered to pay a fine. He committed suicide soon thereafter. The airline, however, had saved its public image by focusing on a single individual who—the court showed—had behaved erratically and unreliably.

Potentially disruptive lessons about the system as a whole are transformed into isolated hick-ups by a few uncharacteristically ill-performing individuals.

This transformation relieves the larger organization of any need to change views and beliefs, or associated policies or spending practices. The system is safe, if only it weren't for a few unreliable humans in it.

FACED WITH A BAD, SURPRISING EVENT, WE CHANGE THE EVENT OR THE PLAYERS IN IT—

RATHER THAN OUR BASIC BELIEFS ABOUT THE SYSTEM THAT MADE THE EVENT POSSIBLE

Instead of modifying our views in the light of the event, we re-shape, re-tell and re-inscribe the event until it fits the traditional and nonthreatening view of the system. As far as organizational learning is concerned, the mishap might as well not have happened. The proximal nature of our reactions to failure makes that expensive organizational lessons can go completely unlearned.

The pilots of a large military helicopter that crashed on a hillside in Scotland in 1994 were found guilty of gross negligence. The pilots did not survive—29 people died in total—so their side of the story could never be heard. The official inquiry had no problems with "destroying the reputation of two good men", as a fellow pilot put it. Indeed, many other pilots felt uneasy about the conclusion. Potentially fundamental vulnerabilities (such as 160 reported cases of Uncommanded Flying Control Movement or UFCM in computerized helicopters alone since 1994) were not looked into seriously.[3]

COUNTERFACTUAL

Finding out what could have prevented the mishap

The outcome of a sequence of events is the starting point of your work as investigator. Otherwise you wouldn't actually be there. This puts you at a remarkable disadvantage when it comes to understanding the point of view of the people you're investigating. Tracing back from the outcome, you will come across joints where people had opportunities to "zig" instead of "zag"; where they could have directed the events away from failure. As investigator you come out on the other end of the sequence of events wondering how people could have missed those opportunities to steer away from failure.

Accident reports are generally full of counterfactuals that describe in fine detail the pathways and options that the people in question did not take. For example, "The airplane could have overcome the windshear encounter if the pitch attitude of 15 degrees nose-up had been maintained, the thrust had been set to 1.93 EPR (Engine Pressure Ratio) and the landing gear had been retracted on schedule."[4]

Counterfactuals prove what could have happened if certain minute and often utopian conditions had been met. Counterfactual reasoning may thus be a fruitful exercise when recommending countermeasures against such failures in the future.

But when it comes to explaining behavior, counterfactuals contribute little. Stressing what was not done (but if it had been done, the accident wouldn't have happened) explains nothing about what actually happened, or why. Counterfactuals are not opportunities missed by the people you are investigating. Counterfactuals are products of your hindsight. Hindsight allows you to transform an uncertain and complex sequence of events into a simple, linear series of obvious options. By stating counterfactuals, you are probably oversimplifying the decision problems faced by people at the time.

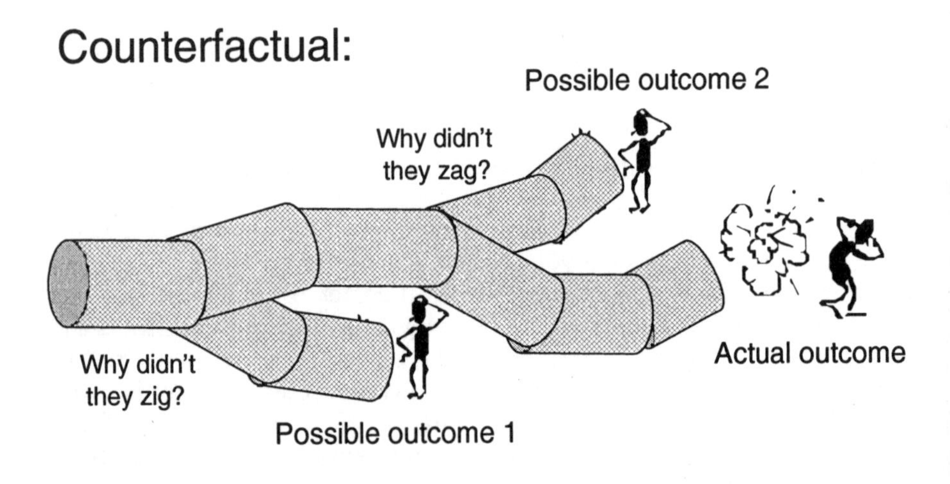

Figure 2.4: Counterfactuals: Going back through a sequence, you wonder why people missed opportunities to direct events away from the eventual outcome. This, however, does not explain failure.

Forks in the road stand out so clearly to you, looking back. But when inside the tunnel, when looking forward and being pushed ahead by unfolding events, these forks were shrouded in the uncertainty and complexity of many possible options and demands; they were surrounded by time constraints and other pressures.

JUDGMENTAL

Saying what they should have done, or failed to do

To explain failure, we seek failure. In order to explain why a failure occurred, we look for errors, for incorrect actions, flawed analyses, inaccurate perceptions. When you have to explain failure, wrong judgments, inaccurate perceptions and missed opportunities would seem like a good place to start.

Yet these decisions, judgments, perceptions are bad or wrong or inaccurate only from hindsight—from your point of view as retrospective outsider. When viewed from the inside of a situation, decisions, judgments and perceptions are just that: decisions, judgments and perceptions. Look at figure 2.5. The very use of the word "failure" in investigative conclusions (for example: "the crew failed to recognize a mode shift") indicates that you are still on the top line, looking down. It represents a judgment from outside the situation, not an explanation from people's point of view within.

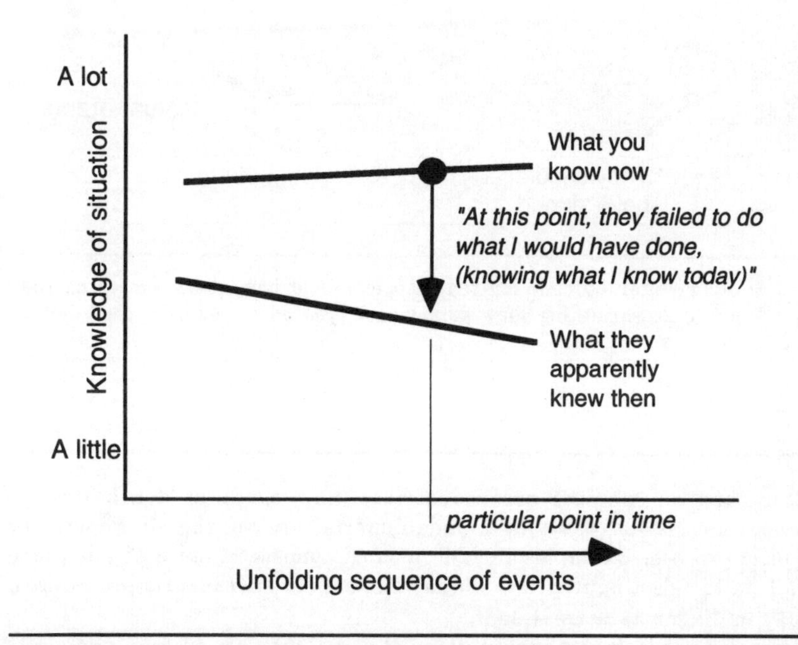

Figure 2.5: Judgmental: saying that other people failed to do what they should have done (knowing what you know today) does not explain their behavior.

The word failure implies an alternative pathway, one which the people in question did not take (for example, recognizing the mode change). Laying out this pathway is counterfactual, as explained above.

But by saying that people "failed" to take this pathway—in hindsight the right one—you judge their behavior according to a standard you can impose only with your broader knowledge of the mishap, its outcome and the circumstances surrounding it. You have not explained a thing yet. You have not shed light on how things looked on the inside of the situation; why people did what they did given *their* circumstances.

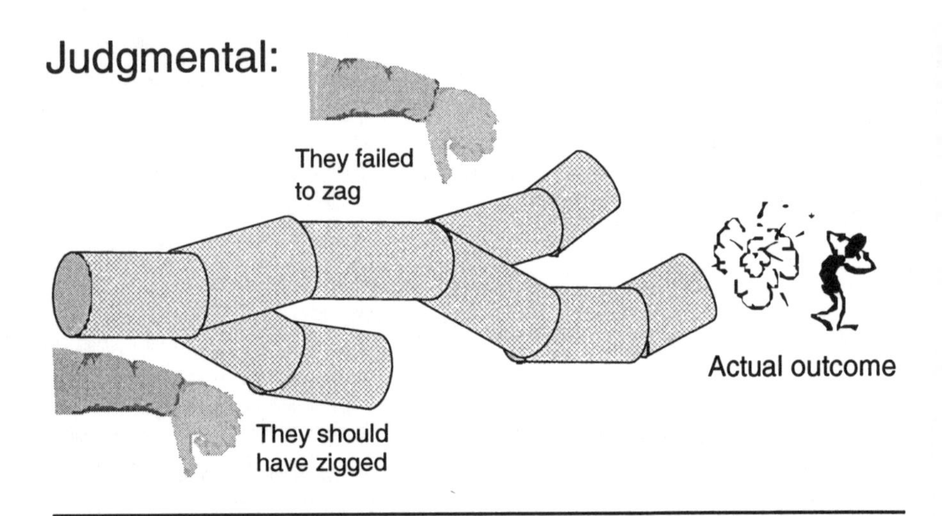

Figure 2.6: Judgmental: by claiming that people should have done something they didn't, or failed to do something they should have, you do not explain their actual behavior.

The literature on medical error describes how cases of death due to negligence may be a result of a judgment failure in the diagnostic or therapeutic process. Examples include a misdiagnosis in spite of adequate data, failure to select appropriate diagnostic tests or therapeutic procedures, and delay in diagnosis or treatment.[5]

Although they look like explanations of error, they are in fact judgments that carry no explanation at all. For example, the "misdiagnosis in spite of adequate data" was once (before hindsight) a reasonable diagnosis based on the data that seemed critical or relevant—otherwise it would not have been made by the physician in question. Calling it a misdiagnosis is an unconstructive, retrospective judgment that misses the reasons behind the actual diagnosis.

The illusion of cause-consequence equivalence

One reason why people feel compelled to judge instead of explain—why they look for failure to explain failure—has to do with "cause-consequence equivalence".

BAD OUTCOME = BAD PROCESS

We assume that really bad consequences can only be the result of really bad causes. Faced with a disastrous outcome, or the potential for one, we assume that the acts leading up to it must have been equally monstrous. Once we know an outcome is bad, we can no longer look objectively at the process that led up to it.

But this automatic response is very problematic in complex worlds. Here even bad processes often lead to good outcomes. And good processes can lead to bad outcomes. Processes may be "bad" in the retrospecitve sense that they departed from routines you now know to have been applicable. But this does not necessarily lead to failure. Given their variability and complexity, these worlds typically offer an envelope of options and pathways to safe outcomes. There is more than one way to success.

BAD PROCESSES MOSTLY LEAD TO GOOD OUTCOMES

GOOD PROCESSES SOMETIMES LEAD TO BAD OUTCOMES

Think of a rushed approach in an aircraft that becomes stabilized at the right time and leads to a safe landing. The opposite goes too. Good processes (in the sense that they do not depart from the drill), where people double-check and communicate and stick to procedures, can lead to disastrous outcomes.

FAILURES AS THE BY-PRODUCT OF NORMAL WORK

What is striking about many accidents is that people were doing exactly the sorts of things they would usually be doing—the things that usually lead to success and safety. People are doing what makes sense given the situational indications, operational pressures and organizational norms existing at the time. Accidents are seldom preceded by bizarre behavior.

If this is the most profound lesson you and your organization can learn from a mishap, it is also the most frightening. The difficulty of accepting this reality lies behind our reactions to failure. Going beyond reacting to failure means acknowledging that failures are baked into the very nature of your work and organization; that they are symptoms of deeper trouble or by-products of systemic brittleness in the way you do your business. It means having to acknowledge that mishaps are the result of everyday influences on everyday decision making, not isolated cases of erratic individuals behaving unrepresentatively. Going beyond your reactions to failure means having to find out why what people did back there and then actually made sense given the organization and operation that surrounded them.

Notes

1 International Herald Tribune, March 14 2001.

2 Wilkinson, S. (1994). The Oscar November Incident. *Air & Space, February-March.*

3 Sunday Times, 25 June 2000.

4 National Transportation Safety Board (1995). *Aircraft Accident Report: Flight into terrain during missed approach USAir flight 1016, DC-9-31, N954VJ, Charlotte, NC, July 2, 1994.* Washington, DC: NTSB, page 119.

5 Bogner, M: S. (Ed.) (1994). *Human error in medicine.* Hillsdale, N.J: Erlbaum.

3. What is the Cause?

What was the cause of the mishap? In the aftermath of failure, no question seems more pressing. There can be significant pressure from all kinds of directions to pinpoint a cause:

- People want to start investing in countermeasures;
- People want to know how to adjust their behavior to avoid the same kind of trouble;
- People may simply seek retribution, punishment, justice.

Two persistent myths drive our search for the cause of failure:

- We think there is something like *the* cause of a mishap (sometimes we call it the root cause, or the primary cause), and if we look in the rubble hard enough, we will find it there. The reality is that there is no such thing as *the* cause, or primary cause or root cause. Cause is something we construct, not find. The first part of this chapter talks about that.
- We think we can make a distinction between human cause and mechanical cause, and that a mishap has to be caused by either one or the other. This is an oversimplification. Once you acknowledge the complexity of pathways to failure, you will find that the distinction between mechanical and human contributions becomes very blurred; even impossible to maintain. The second part of this chapter talks about that.

THE CONSTRUCTION OF CAUSE

Look at two official investigations into the same accident. One was conducted by the airline whose aircraft crashed somewhere in the mountains. The other was conducted by the civil aviation authority of the country in which the accident occurred, and who employed the air traffic controller in whose airspace it took place.

The authority says that the controller did not contribute to the cause of the accident, yet the airline claims that air traffic control clearances were not in accordance with applicable standards and that the controller's inadequate language skills and inattention were causal. The authority counters that the pilot's inadequate use of flightdeck automation was actually to blame, whereupon the airline points to an inadequate navigational database supplied to their flight computers among the causes. The authority explains that the accident was due to a lack of situation awareness regarding terrain and navigation aids, whereas the airline blames lack of radar coverage over the area. The authority states that the crew failed to revert to basic navigation when flight deck automation usage created confusion and workload, whereupon the airline argues that manufacturers and vendors of flightdeck automation exuded overconfidence in the capabilities of their technologies and passed this on to pilots. The authority finally blames ongoing efforts by the flight crew to expedite their approach to the airport in order to avoid delays, whereupon the airline lays it on the controller for suddenly inundating the flight crew with a novel arrival route and different runway for landing.[1]

Table 3.1: Two statements of cause about the same accident

Causes according to Authority:	Causes according to Airline:
Air Traffic Controller did not play a role	No standard phraseology, inadequate language and inattention by Controller
Pilots' inadequate use of automation	Inadequate automation database
Loss of pilots' situation awareness	Lack of radar coverage over area
Failure to revert to basic navigation	Overconfidence in automation sponsored by vendors
Efforts to hasten arrival	Workload increase because of Controller's sudden request

So who is right? The reality behind the controversy, of course, is that both investigations are right. They are both right in that all of the factors mentioned were in some sense causal, or contributory, or at least necessary. Make any one of these factors go away and the sequence of events would probably have turned out differently. But this also means that both sets of claims are wrong. They are both wrong in that they focus on only a subset of contributory factors and pick and choose which ones are causal and which ones are not. This choosing can be driven more by socio-political and organizational pressures than by mere evidence found in the rubble. Cause is not something you find. Cause is something you construct. How you construct it and from what evidence, depends on where you look, what you look for, who you talk to, what you have seen before, and likely on who you work for.

There is no "root" or "primary" cause

How is it that a mishap gives you so many causes to choose from? This has to do with the fact that the kinds of systems that are vulnerable to human error are so well protected against it. The potential for danger in many industries and systems has been recognized long ago. Consequently, major investments have been made in protecting them against the breakdowns that we know or think can occur. These so-called "defenses" against failure contain human and engineered and organizational elements.

Flying the right approach speeds for landing while an aircraft goes through its subsequent configurations (of flaps and slats and wheels that come out), is safety-critical. As a result it has evolved into a well-defended process of double-checking and cross-referencing between crew members, speed booklets, aircraft weight, instrument settings, reminders and call-outs, and in some aircraft even by engineered interlocks.

Accidents in such systems can occur only if multiple factors succeed in eroding or bypassing all these layers of defense. The breach of any of these layers can be called "causal". For example, the crew opened the speed booklet on the wrong page (i.e. the wrong aircraft landing weight). But this fails to explain the entire breakdown, because other layers of defense had to be broken or side-stepped too. And there is another question. Why did the crew open the booklet on the wrong page? In other words, what is the cause of that action? Was it their expectation of aircraft weight based on fuel used on that trip; was it a misreading of an instrument? And once pinpointed, what is the cause of that cause? And so forth.

Because of this investment in multiple layers of defense, we can find "causes" of failures everywhere—when they happen, that is. The causal web quickly multiplies and fans out, like cracks in a window. What you call "root cause" is simply the place where you stop looking any further. As far as the causal web is concerned, there are no such things as root or primary causes—there is in fact no end anywhere. If you find a root or primary cause, it was your decision to distinguish something in the dense causal pattern by those labels.

There is no single cause

So what is the cause of the accident? This question is just as bizarre as asking what *the* cause is of not having an accident. There is no single cause. Neither for failure, nor for success. In order to push a well-defended system over the edge (or make it work safely), a large number of contributory factors are necessary and only jointly sufficient.

MULTIPLE FACTORS—EACH NECESSARY AND ONLY JOINTLY SUFFICIENT—ARE NEEDED TO PUSH A COMPLEX SYSTEM OVER THE EDGE OF BREAKDOWN

So where you focus in your search for cause is something that the evidence in a mishap will not necessarily determine for you. It is up to your investigation.

In a break with the tradition of identifying "probable causes" in aviation crashes—which oversimplify the long and intertwined pathway to failure—Judge Moshansky's investigation of the Air Ontario crash at Dryden, Canada in 1989 did not produce any probable causes. The pilot in question had made a decision to take off with ice and snow on the wings, but, as Moshanky's commission wrote, "that decision was not made in isolation. It was made in the context of an integrated air transportation system that, if it had been functioning properly, should have prevented the decision to take off...there were significant failures, most of them beyond the captain's control, that had an operational impact on the events at Dryden...regulatory, organizational, physical and crew components...."

Instead of forcing this complexity into a number of probable causes, the Commission generated 191 recommendations which pointed to the many "causes" or systemic failures underlying the symptomatic accident on that day in March 1989. Recommendations ranged in topic from the introduction of a new aircraft type to a fleet, to management selection and turn-over in the airline, to corporate take-overs and mergers in the aviation industry.[2]

Probable cause statements are of necessity:

- Selective. There are only so many things you can label "causal" before the word "causal" becomes meaningless.;
- Exclusive. They leave out factors that were also necessary and only jointly sufficient to "cause" the failure;
- Oversimplifications. They highlight only a few hotspots along a long, twisted and highly interconnected causal pathway that starts long before and far away from where the actual failure occurs.

If protocol prescribes that probable causes be identified, the best way to deal with that is to generate, as "probable cause", the shortest possible summary of the sequence of events that led up to the mishap (see also chapter 10). This description should start as high up in the causal chain as possible, and follow the meandering pathway to the eventual failure. Although an oversimplification, this minimizes selectivity and exclusion by highlighting points all along a causal network.

HUMAN OR SYSTEM FAILURE?

Was this mishap due to human error, or did something else in the system break? You hear the question over and over again—in fact, it is often the first question people ask. The question, however, demonstrates an oversimplified belief in the roots of failure. And it only very thinly disguises the bad apple theory: the system is basically safe, but it contains unreliable components. These components are either human or mechanical, and if one of them fails, a mishap ensues.

Early investigation can typically show that a system, for example an aircraft, behaved as designed or programmed on its way into a mountainside, and that there was nothing mechanically wrong with it. This is taken as automatic evidence that the mishap must have been caused by human error—after all, nothing was wrong with the system. If this is the conclusion, then the old view of human error has prevailed. Human error causes failure in otherwise safe, well-functioning systems. The reality, however, is that the "human error" does not come out of the blue. Error has its roots in the system surrounding it; connecting systematically to mechanical, programmed, paper-based, procedural, organizational and other aspects to such an extent that the contributions from system and human begin to blur.

Passenger aircraft have "spoilers"—panels that come up from the wing upon landing, to help brake the aircraft during its landing roll-out. To make these spoilers come out, pilots have to manually "arm" them by pulling a lever in the cockpit. Many aircraft have landed without the spoilers being armed, some cases even resulting in runway overruns. Each of these events gets classified as "human error"—after all, the human pilots forgot something in a system that is functioning perfectly otherwise.

But deeper probing reveals a system that is not at all functioning perfectly. Spoilers typically have to be armed after the landing gear has come out and is safely locked into place. The reason is that landing gears have compression switches which communicate to the aircraft when it is on the ground. When the gear compresses, it tells the aircraft that it has landed. And then the spoilers come out (if they are armed, that is). Gear compression, however, can also occur *while* the gear is coming out, because of airpressure from the slip stream around a flying aircraft, especially if landing gears fold open *into* the wind. This would create a case where the aircraft thinks it is on the ground, when it is not. If the spoilers would already be armed at that time, they would come out too—not good while still airborne. To prevent this, aircraft carry procedures for the spoilers to be armed only when the gear is fully down and locked. It is safe to do so, because the gear is then orthogonal to the slipstream, with no more risk of compression.

But the older an aircraft gets, the longer a gear takes to come out and lock into place. The hydraulic system no longer works as well, for example. In some aircraft, it can take up to half a minute. By that time, the gear extension has begun to intrude into other cockpit tasks that need to happen—selecting wing flaps for landing; capturing and tracking the electronic glide slope towards the runway; and so forth. These are items that come *after* the "arm spoilers' item on a typical before-landing checklist. If the gear is still extending, while the world has already pushed the flight further down the checklist, not arming the spoilers is a slip that is easy to make.

Combine this with a system that, in many aircraft, never warns pilots that their spoilers are not armed; a spoiler handle that sits over to one, dark side of the center cockpit console, obscured for one pilot by power levers, and whose difference between armed and not-armed may be all of one inch, and the question becomes: is this mechanical failure or human error?

One pilot told me how he, after years of experience on a particular aircraft type, figured out that he could safely arm the spoilers 4 seconds after "gear down" was selected, since the critical time for potential gear compression was over by then. He had refined a practice whereby his hand would go from the gear lever to the spoiler handle slowly enough to cover 4 seconds—but it would always travel there first. He thus bought enough time to devote to subsequent tasks such as selecting landing flaps and capturing the glide slope. This is how practitioners create safety: they invest in their understanding of how systems can break down, and then devise strategies that help forestall failure.

The deeper you dig, the more you will understand why people did what they did, based on the tools and tasks and environment that surrounded them. The further you push on into the territory where their errors came from, the more you will discover that the distinction between human and system failure does not hold up.

Statistics and the 70% myth

The supposed distinction between human error and system failure also creates the idea that we can statistically tabulate each and get a quick overview of the chief safety challenges to our systems. Surprise! Human error always wins. The assertion that at least 70% of mishaps are due to human error is particularly stable, and consistent across industries. It gives the bad news about system safety both a concrete source and a number. The bad apple theory has become quantified.

Tabulation of errors may have worked once upon a time, when tightly controlled laboratory studies were set up to investigate human performance. In these lab studies, human tasks and opportunities to err were shrunk to a bare minimum, and singular, measurable errors could be counted as a basic unit of human performance. This kind of experimentation left the scientist with spartan but quantifiable results. Yet when it comes to human error "in the wild"—that is, as it occurs in natural complex settings—such tabulation and percentages obscure many things and muffles learning from failure:

- Percentages ignore the fact that complex interactions between human and various other contributions are typically necessary to move a system towards breakdown. These 70% human errors do not occur as erratic slips or brain bloopers in the vacuum of a perfectly engineered or rationally organized world. In real tales of failure, the actions and assessments we call "errors" are intermixed with problems of many other kinds: mechanical, organizational. The bad news lies not in the 70% human errors, but in the interactions between human behavior and features and vulnerabilities of their operating worlds.

- Percentages hide the wide diversity of human error in the wild. As symptoms of deeper problems, the expression of human error is context-dependent. The kind of error is determined in large part by features of the circumstances in which it takes place. The details of why tasks and tools and working environments are vulnerable to errors—or why they may even invite a large percentage of errors in the first place—get lost under the large label of "human error".

There are additional problems with the 70% myth. For example, what do we refer to when we say "error"? In safety debates there are three ways of using the label "error":

- Error as the *cause* of failure. For example: This event was due to human error.
- Error as the *failure itself*. For example: The pilot's selection of that mode was an error.
- Error as a *process*, or, more specifically, as a departure from some kind of standard. This may be operating procedures, or simply good airmanship. Depending on what you use as standard, you will come to different conclusions about what is an error.

So which one do people actually adhere to when they categorize mishaps according to "human error"? Defining human error as cause is completely unproductive. The myth is that 70% represents the distance we have to go before we reach full safety. Full safety lies somewhere on, or beyond, the horizon, and the 70% human errors is what is between us and that goal. This assumption about the location of safety is an illusion, and efforts to measure the distance to it are little more than measuring our distance from a mirage.

Safety is right here, right now, right under our feet—not yonder across some 70%. Look back at the spoiler example above. People in complex systems create safety. They make it their job to anticipate forms of, and pathways toward, failure, they invest in their own resilience and that of their system by tailoring their tasks, by inserting buffers, routines, heuristics, tricks, double-checking, memory aids. The

70% human contribution to failure occurs because complex systems need an overwhelming human contribution for their safety. Human error is the inevitable by-product of the pursuit of success in an imperfect, unstable, resource-constrained world. To try to eradicate human error (to depress or reduce the 70%) would mean to eradicate or compromise human expertise—the most profound and most reliable investment in system safety and success we could ever hope for. In order to understand the 70% human error, we need to understand the 70% (or more) contribution that human expertise makes to system success and safety.

Notes

1 See: Aeronautica Civil (1996). *Aircraft Accident Report: Controlled flight into terrain American Airlines flight 965, Boeing 757-223, N851AA near Cali, Colombia, December 20, 1995*. Santafe de Bogota, Colombia: Aeronautica Civil Unidad Administrativa Especial, and American Airlines' (1996) Submission to the Cali Accident Investigation.

2 Moshansky, V. P. (1992). *Commission of inquiry into the Air Ontario accident at Dryden, Ontario* (Final report, vol. 1-4). Ottawa, ON: Minister of Supply and Services, Canada.

4. Human Error by any Other Name

"A spokesman for the Kennedy family has declined to comment on reports that a federal investigation has concluded that pilot error caused the plane crash that killed John F. Kennedy Jr., his wife and his sister-in-law. The National Transportation Safety Board is expected to finish its report on last year's crash and release it in the next several weeks. Rather than use the words 'pilot error', however, the safety board will probably attribute the cause to Kennedy's becoming 'spatially disoriented', which is when a pilot loses track of the plane's position in the sky."[1]

UNDERSPECIFIED LABELS

"Human error" as explanation for accidents has become increasingly unsatisfying. As mentioned earlier, there is always an organizational world that lays the groundwork for the "errors", and an operational one that allows them to spin into larger trouble.

We also know there is a psychological world behind the errors—to do with people's attention, perception, decision making, and so forth. Human factors has produced or loaned a number of terms that try to capture such phenomena. Labels like "complacency", "situation awareness", "crew resource management", "shared mental models", "stress", "workload", "non-compliance" are such common currency today that nobody really dares to ask what they actually mean. The labels are assumed to speak for themselves; to be inherently meaningful. They get used freely as causes to explain failure. For example:

- "The crew lost situation awareness and effective crew resource management (CRM)" (which is why they crashed);
- "High workload led to a stressful situation" (which is why they got into this incident);

- "It is essential in the battle against complacency that crews retain their situation awareness" (otherwise they keep missing those red signals).
- "Non-compliance with procedures is the single largest cause of human error and failure" (so people should just follow the rules).

The question is: are labels such as complacency or situation awareness much better than the label "human error"? In one sense they are. They provide some specification; they appear to give some kind of reasons behind the behavior; they provide an idea of the sort of circumstances and manner in which the error manifested itself.

But if they are used as quoted above, they do not differ from the verdict "human error" they were meant to replace. These labels actually all conclude that human error—by different names—was the cause:

- Loss of CRM is one name for human error—the failure to invest in common ground, to coordinate operationally significant data among crewmembers;
- Loss of situation awareness is another name for human error—the failure to notice things that in hindsight turned out to be critical;
- Complacency is also a name for human error—the failure to recognize the gravity of a situation or to follow procedures or standards of good practice.
- Non-compliance is also a name for human error—the failure to stick with standard procedures that would keep the job safe.

Dealing with the illusion of explanation

Human factors risks falling into the trap of citing "human error" by any other name. Just like "human error", labels like the ones above also hide what really went on, and instead simply judge people for what they did not do (follow the rules, coordinate with one another, notice a signal, etc.). Rather than explaining why people did what they did, these labels say "human error" over and over again. They get us nowhere. As shown in chapter 2, judging people is easy. Explaining why their assessments and actions made sense is hard. The labels discussed in this chapter may give you the illusion of explanation, but they are really judgments in disguise. Saying that other people lost situation awareness, for example, is really saying that you now know more about their situation than they did back then, and then you call it their error. This of course explains nothing.

Here is why labels like "complacency" or "loss of situation awareness" must not be mistaken for deeper insight into human factors issues. The risk occurs when they are applied by investigators without making explicit connections between:

- The label and the evidence for it. For example, exactly which interactions and miscoordinations in the sequence of events constituted a loss of effective crew resource management—based on available and accepted models of "effective crew resource management"?
- The label and how it "caused" the mishap. For example, "loss of effective crew resource management" may be cited in the probable causes or conclusions. But how exactly did the behaviors that constituted its loss contribute to the outcome of the sequence of events?

If you reveal which kinds of behaviors in the sequence of events produced a "loss of effective crew resource management", these behaviors can themselves point to the outcome, without you having to rely on a label that obscures all the interesting bits and interactions.

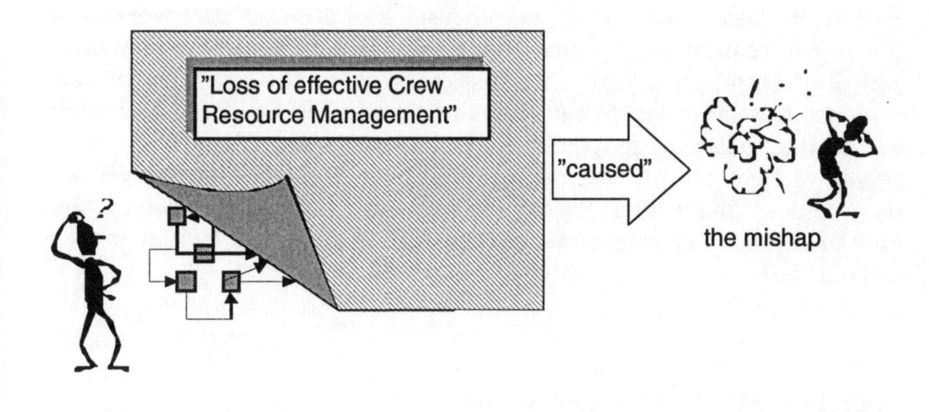

Figure 4.1: The interesting mental dynamics take place *beneath* the large psychological label. The label itself explains nothing.

To understand the mindset of someone caught up in an unfolding situation is not a matter of translating his or her behavior into big psychological terms. It's the mental dynamics *beneath* the labels that are interesting—for example:

- The ways people shift attention on the basis of earlier assessments of the situation or on the basis of future expectations;
- The trade-offs they have to make between various operational or organizational goals;
- How they activate and apply knowledge in context;
- How they recognize patterns of data on the basis of experience with similar circumstances;
- How they coordinate with various sources of expertise inside and outside their situation;
- How they deal with the clumsiness and complexity of the technology that surrounds them.

Chapters 9 and 10 will help you with deeper insight into these issues. It's these mental and interpersonal processes that drive a sequence of events in certain directions; it's these processes—if anything—that can be said to be "causal" in the sense that they help determine the outcome of a sequence of events.

The use of large terms in investigative findings and explanations may be seen as the rite of passage into psychological phenomena. That is, for a human factors investigation to be taken seriously, it should contain its dose of situation awarenesses and stresses and workloads. But in the rush to come across like a solid human factors investigator, you may forget that you can't just jump from the specifics in your evidence to a large label that seems to cover it all. You need to explain something in between; you need to leave a trace. Otherwise other people will get completely lost and will have no idea whether you are right or not. This is what we call "folk science", or "folk psychology" and human factors investigations can be full of it. Let us look at it for a little bit here.

FOLK SCIENCE IN HUMAN FACTORS

How do you know whether something is a folk model, as opposed to a human factors concept that actually has some scientific merit? Here is a rough guide:

- Folk models substitute one big term for another instead of defining the big term by breaking it down into more little ones (in science we call this decomposition, or deconstruction).

- Folk models are difficult to prove wrong, because they do not have a definition in terms of smaller components, that are observable in people's real behavior. Folk models may seem glib; they appeal to popular understandings of difficult phenomena.
- Folk models easily lead to overgeneralization. Before you know it, you may see "complacency" and "loss of situation awareness" everywhere. This is possible because the concepts are so ill-defined. You are not bound to particular definitions, so you may interpret the concepts any way you like.

Take as an example an automation-related accident that occurred when situation awareness or automation-induced complacency did not yet exist—in 1973. The issue was an aircraft on approach in rapidly changing weather conditions that was equipped with a slightly deficient "flight director" (a device on the central instrument showing the pilot where to go, based on an unseen variety of sensory inputs), and which the captain of the airplane in question distrusted. The airplane struck a seawall bounding Boston's Logan airport about a kilometer short of the runway and slightly to the side of it, killing all 89 people onboard. In its comment on the crash, the transport safety board explained how an accumulation of discrepancies, none critical in themselves, can rapidly deteriorate, without positive flight management, into a high-risk situation. The first officer, who was flying, was preoccupied with the information presented by his flight director systems, to the detriment of his attention to altitude, heading and airspeed control.

Today, both automation-induced complacency on part of the first officer and a loss of situation awareness on part of the entire crew would most likely be cited under the causes of this crash. (Actually, that the same set of empirical phenomena can comfortably be grouped under either label (complacency or loss of situation awareness) is additional testimony to the undifferentiated and underspecified nature of these human factors concepts). These "explanations" (complacency, loss of situation awareness) were obviously not necessary in 1973 to deal with this accident. The analysis left us instead with more detailed, more falsifiable, and more traceable assertions that linked features of the situation (e.g. an accumulation of discrepancies) with measureable or demonstrable aspects of human performance (diversion of attention to the flight director versus other sources of data). The decrease in falsifiability represented by complacency and situation awareness as hypothetical contenders in explaining this crash is really the inverse of scientific progress. Promise derives from being better than what went before—which these models, and the way in which they would be used, are not.

Chapter 10 may point you to a more productive avenue. By going through various typical patterns of failure, it discusses human factors concepts that have more scientific merit in the sense that they are supported by better articulated models of human performance. Using this, you may find it easier to recognize some of these patterns of failure in the sequence of events you are investigating. More importantly, you may find it easier to leave a trace for others to follow, to show them your analysis, to make them understand why you came to the conclusions that you drew.

Note

1 International Herald Tribune, 24-25 June 2000.

5. Human Error—in the Head or in the World?

The use of underspecified labels in human error investigations, covered in the previous chapter, has various roots. One reason for the use of large psychological terms is the confusion over whether you should start looking for the source of human error:

- In the head (of the person committing the error)
- Or in the situation (in which the person works).

The first alternative is used in various human error analysis tools, and in fact often implied in investigations. For example, when you use "complacency" as a label to explain behavior, you really look for how the problem started with an individual who was not sufficiently motivated to look closely at critical details of his or her situation.

As said in the previous chapters, such an approach to "explaining" human error is a dead-end. It prevents an investigation from finding enduring features of the operational environment that actually produce the controversial behavior (and that will keep producing it if left in place). And there is more. The assumption that errors start in the head also leaves an investigative conclusion hard to verify for others, as is explained below.

The alternative—look for the source of error in the world—is a more hopeful path for investigations. Human error is systematically linked to features of the world—the tasks and tools that people work with, and the operational and organizational environment in which people carry out that work. If you start with the situation, you can identify, probe and document the reasons for the observed behavior, without any need to resort to non-observable processes or structures or big labels in someone's head. This is the path that The Field Guide will take you along.

HUMAN ERROR—IS IT ALL IN THE HEAD?

To "reverse engineer" human error, chapter 9 will encourage you to reconstruct how people's mindset unfolded and changed over time. You would think that reconstructing someone's unfolding mindset begins with the mind. The mind, after all, is the obvious place to look for the mindset that developed inside of it. Was there a problem holding things in working memory? What was in the person's perceptual store? Was there trouble retrieving a piece of knowledge from long-term memory? These are indeed the kinds of questions asked in a variety of human error analysis tools and incident reporting systems.

A tool is being developed for the analysis of human errors in air traffic control. For each observed error, it takes the analyst through a long series of questions that are based on an elaborate information processing model of the human mind. It begins with perceptual processes and points the analyst to possible problems or difficulties there. Then it goes on along the processing pathway, hoping to guide the analyst to the source of trouble in a long range of psychological processes or structures: short term memory, long term memory, decision making, response selection, response execution, and even the controller's image of him or herself. For each observed error, the journey through the questions can be long and arduous and the final destination (the supposed source of error) dubious and hard to verify.

These human error analyses deal with the complexity of behavior by simplifying it down to boxes; by nailing the error down to a single psychological process or structure. For example, it was an error of perceptual store, or one of working memory, or one of judgment or decision making, or one of response selection. The aim is to conclude that the error originated in a certain stage along a psychological processing pathway in the head. These approaches basically explain error by taking it back to the mind from which it came.

The shortcomings, as far as investigating human error is concerned, are severe. These approaches hide an error back in the mind under a label that is not much more enlightening than "human error" is. In addition, the labels made popular in these approaches (such as working memory or response execution) are artifacts of the language of a particular psychological model. This model may not even be right, but it sure is hard to prove wrong. Who can prove the existence of short term memory? But who can prove that it does not exist?

Explaining human error on the basis of internal mental structures will leave other people guessing as to whether the investigator was right or not. Nobody can actually see things like short term memories or perceptual stores, and nobody can go back into the short term memories or perceptual stores of the people involved to check the investigator's work. Other people can only hope the investigator was right when the psychological category was picked.

By just relabeling human error in more detailed psychological terms, investigations remain locked in a practice where anyone can make seemingly justifiable, yet unverifiable assertions. Such investigations remain fuzzy and uncertain and inconclusive, and low on credibility.

HUMAN ERROR—MATTER OVER MIND

Things are different when you begin your investigation with the unfolding situation in which people found themselves. Methods that attribute human error to structures inside the brain easily ignore the situation in which human behavior took place, or they at least underestimate its importance. Yet it makes sense to start with the situation:

- Past situations can be objectively reconstructed to a great extent, and documented in detail;
- There are tight and systematic connections between situations and behavior; between what people did and what happened in the world around them.

These connections between situations and behavior work both ways:

- People change the situation by doing what they do; by managing their processes;
- But the evolving situation also changes people's behavior. An evolving situation provides changing and new evidence; it updates people's understanding; it presents more difficulties; it forecloses or opens pathways to recovery.

You can uncover the connections between situation and behavior, investigate them, document them, describe them, represent them graphically. Other people can look at the reconstructed situation and how you related it to the behavior that took place inside of it. Other people can actually trace your explanations and conclusions. Starting with the situation brings a human error investigation out in the open. It does not rely on hidden psychological structures or processes, but instead allows verification and debate by those who understand the domain. When a human error investigation starts with the situation, it sponsors its own credibility.

A large part of human error investigations, then, is not at all about the human behind the error. It is not about supposed structures in a human's mind; about psychological constructs that were putatively involved in causing mental hick-ups. A large part of human error investigation is about the situation in which the human was working; about the tasks he or she was carrying out; about the tools that were used.

THE RECONSTRUCTION OF MINDSET BEGINS NOT WITH THE MIND

IT BEGINS WITH THE CIRCUMSTANCES IN WHICH THE MIND FOUND ITSELF

To understand the situation that produced and accompanied behavior, is to understand the human assessments and actions inside that situation. This allows you to "reverse engineer" human error by showing:

- how the process, the situation, changed over time;
- how people's assessments and actions evolved in parallel with their changing situation;
- how features of people's tools and tasks and their organizational and operational environment influenced their assessments and actions inside that situation.

This is what the reconstruction of unfolding mindset, the topic of chapter 9, is all about.

6. Put Data in Context

Putting behavior back into the situation that produced and accompanied it is not easy. In fact, to make sense of behavior it is always tempting to go for a context that actually lies *outside* the mishap sequence. Taking behavior out of context, and giving it meaning from the outside, is common in investigations. This chapter discusses two ways in which behavioral data is typically taken out of context, by:

- micro-matching them with a world you now know to be true, or by
- lumping selected bits together under one condition you have identified in hindsight ("cherry picking").

OUT OF CONTEXT I:
HOLDING PERFORMANCE FRAGMENTS AGAINST A WORLD YOU NOW KNOW TO BE TRUE

One of the most popular ways by which investigators assess behavior is to hold it up against a world they *now* know to be true. There are various ways in which after-the-fact-worlds can be brought to life:

- A procedure or collection of rules: People's behavior was not in accordance with standard operating procedures that were found to be applicable for that situation afterward;
- A set of cues: People missed cues or data that turned out to be critical for understanding the true nature of the situation;
- Standards of good practice: People's behavior fall short of standards of good practice in the particular industry.

The problem is that these after-the-fact-worlds may have very little in common with the actual world that produced the behavior under investigation. They contrast people's behavior against the investigator's reality, not the reality that surrounded the behavior in question. Thus, micro-matching fragments of behavior with these various standards explains nothing—it only judges.

Procedures

First, individual fragments of behavior are frequently compared with procedures or regulations, which can be found to have been applicable in hindsight. Compared with such written guidance, actual performance is often found wanting; it does not live up to procedures or regulations.

Take the automated airliner that started to turn towards mountains because of a computer-database anomaly. The aircraft ended up crashing in the mountains. The accident report explains that one of the pilots executed a computer entry without having verified that it was the correct selection, and without having first obtained approval of the other pilot, contrary to the airline's procedures.[1]

Investigations invest considerably in organizational archeology so that they can construct the regulatory or procedural framework within which the operations took place or should have taken place. Inconsistencies between existing procedures or regulations and actual behavior are easy to expose in hindsight. Your starting point is a fragment of behavior, and you have the luxury of time and resources to excavate organizational records and regulations to find rules with which the fragment did not match.

But what have you shown? You have only pointed out that there was a mismatch between a fragment of human performance and existing guidance that you uncovered or highlighted after-the-fact. This is not very informative. Showing that there was a mismatch between procedure and practice sheds little light on the *why* of the behavior in question. And, for that matter, it sheds little light on the why of this particular mishap. Mismatches between procedure and practice are not unique ingredients of accident sequences. They are often a feature of daily operational life (which is where the interesting bit in your investigation starts).

Available data

Second, to construct the world against which to evaluate individual performance fragments, investigators can turn to data in the situation that were not noticed but that, in hindsight, turned out to be critical.

Continue with the automated aircraft above. What should the crew have seen in order to notice the turn? They had plenty of indications, according to the manufacturer of their aircraft:

"Indications that the airplane was in a left turn would have included the following: the EHSI (Electronic Horizontal Situation Indicator) Map Display (if selected) with a curved path leading away from the intended direction of flight; the EHSI VOR display, with the CDI (Course Deviation Indicator) displaced to the right, indicating the airplane was left of the direct Cali VOR course, the EaDI indicating approximately 16 degrees of bank, and all heading indicators moving to the right. Additionally the crew may have tuned Rozo in the ADF and may have had bearing pointer information to Rozo NDB on the RMDI".[2]

This is a standard response after mishaps: point to the data that would have revealed the true nature of the situation. But knowledge of the "critical" data comes only with the privilege of hindsight. If such critical data can be shown to have been physically available, it is automatically assumed that it should have been picked up by the operators in the situation. Pointing out, however, that it should have been does not explain why it was perhaps not, or why it was interpreted differently back then. There is a difference between:

- **Data availability**: what can be shown to have been physically available somewhere in the situation
- **Data observability**: what would have been observable given the features of the interface and the multiple interleaving tasks, goals, interests, knowledge and even culture of the people looking at it.

The mystery, as far as an investigation is concerned, is not why people could have been so unmotivated or stupid not to pick up the things that you can decide were critical in hindsight. The mystery is to find out what *was* important to them, and why.

Other standards

Third, there are a number of other standards especially for performance fragments that do not easily match procedural guidance or for which it is more difficult to point out data that existed in the world and should have picked up. This is often the case when a controversial fragment knows no clear pre-ordained guidance but relies on local, situated judgment. For example, a decision to accept a runway change, or continue flying into bad weather. For these cases there are always "standards of good practice" which are based on convention and putatively practiced across an entire industry. One such standard in aviation is "good airmanship", which, if nothing else can, will cover the variance in behavior that had not yet been accounted for.

Cases for medical negligence can often be made only by contrasting actual physician performance against standards of proper care or good practice. Rigid, algorithmic procedures generally cannot live up to the complexity of the work and the ambiguous, ill-defined situations in which it needs to be carried out. Consequently, it cannot easily be claimed that this or that checklist should have been followed in this or that situation.

But which standards of proper care do you invoke to contrast actual behavior against? This is largely arbitrary, and driven by hindsight. After wrong-site surgery, for example, the standard of good care that gets invoked is that physicians have to make sure that the correct limb is amputated or operated upon. Finding vague or broad standards in hindsight does nothing to elucidate the actual circumstances and systemic vulnerabilities which in the end allowed wrong-site surgery to take place.

By referring to procedures, physically available data or standards of good practice, investigators can micro-match controversial fragments of behavior with standards that seem applicable from their after-the-fact position. Referent worlds are constructed from outside the accident sequence, based on data investigators now have access to, based on facts they now know to be true. The problem is that these after-the-fact-worlds may have very little relevance to the circumstances of the accident sequence. They do not explain the observed behavior. The investigator has substituted his own world for the one that surrounded the people in question.

OUT OF CONTEXT II: GROUPING SIMILAR PERFORMANCE FRAGMENTS UNDER A LABEL IDENTIFIED IN HINDSIGHT

The second way in which data are taken out of context; in which they are given meaning from the outside, is by grouping and labeling behavior fragments that appear to represent a common condition.

Consider this example, where diverse fragments of behavior are lumped together to build a case for haste as explanation of the bad decisions taken by the crew. The fragments are actually not temporally co-located. They are spread out over a considerable time, but that does not matter. According to the investigation they point to a common condition.

"Investigators were able to identify a series of errors that initiated with the flightcrew's acceptance of the controller's offer to land on runway 19...The CVR indicates that the decision to accept the offer to land on runway 19 was made jointly by the captain and the first officer in a 4-second exchange that began at 2136:38. The captain asked: 'would you like to shoot the one nine straight in?' The first officer responded, 'Yeah, we'll have to scramble to get down. We can do it.' This interchange followed an earlier discussion in which the captain indicated to the first officer his desire to hurry the arrival into Cali, following the delay on departure from Miami, in an apparent effort to minimize the effect of the delay on the flight attendants' rest requirements. For example, at 2126:01, he asked the first officer to 'keep the speed up in the descent'... The evidence of the hurried nature of the tasks performed and the inadequate review of critical information between the time of the flightcrew's acceptance of the offer to land on runway 19 and the flight's crossing the initial approach fix, ULQ, indicates that insufficient time was available to fully or effectively carry out these actions. Consequently, several necessary steps were performed improperly or not at all". (Aeronautica Civil, 1996, p. 29)

As one result of the runway change and self-imposed workload the flight crew also "lacks situation awareness"—an argument that is also constructed by grouping voice utterance fragments from here and there:

"...from the beginning of their attempt to land on runway 19, the crew exhibited a lack of awareness.... The first officer asked 'where are we', followed by 'so you want a left turn back to ULQ. The captain replied, 'hell no, let's press on to... and the first officer stated 'well, press on to where though?'.... Deficient situation awareness is also evident from the captain's interaction with the Cali air traffic controller".[3]

It is easy to pick through the evidence of an accident sequence and look for fragments that all seem to point to a common condition. The investigator treats the voice record as if it were a public quarry to select stones from, and the accident explanation the building he needs to construct from those stones. Among investigators this practice is sometimes called "cherry picking"—selecting those bits that help their *a-priori* argument. The problems associated with cherry picking are many:

- You probably miss all kinds of details that are relevant to explaining the behavior in question;
- Each cherry, each fragment, is meaningless outside the context that produced it. Each of the bits that gets lumped together with other "similar" ones actually has its own story, its own background, its own context and its own reasons for being. When it was produced it may have had nothing to do with the other fragments it is now grouped with. The similarity is entirely in the eye of the retrospective beholder.
- Much performance, much behavior, takes place *in between* the fragments that the investigator selects to build his case. These intermediary episodes contain changes and evolutions in perceptions and assessments that separate the excised fragments not only in time, but also in meaning.

Thus, the condition that binds similar performance fragments together has little to do with the circumstances that brought each of the fragments forth; it is not a feature of those circumstances. It is an artifact of you as investigator. The danger is that you come up with a theory that guides the search for evidence about itself. This leaves your investigation not with findings, but with tautologies. What is the solution?

PUT DATA IN CONTEXT

Taking data out of context, either by:

- micro-matching them with a world you now know to be true, or by
- lumping selected bits together under one condition identified in hindsight

robs data of its original meaning. And these data out of context are simultaneously given a new meaning—imposed from the outside and from hindsight. You impose this new meaning when you look at the data in a context you *now* know to be true. Or you impose meaning by tagging an outside label on a loose collection of seemingly similar fragments.

To understand the actual meaning that data had at the time and place it was produced, you need to step into the past yourself. When left or relocated in the context that produced and surrounded it, human behavior is inherently meaningful.

Historian Barbara Tuchman put it this way: "Every scripture is entitled to be read in the light of the circumstances that brought it forth. To understand the choices open to people of another time, one must limit oneself to what they knew; see the past in its own clothes, as it were, not in ours."[4]

Notes

1 The accident report is: Aeronautica Civil (1996). *Aircraft Accident Report: Controlled flight into terrain American Airlines flight 965, Boeing 757-223, N851AA near Cali, Colombia, December 20, 1995*. Santafe de Bogota, Colombia: Aeronautica Civil Unidad Administrativa Especial.
2 Boeing submission to the American Airlines Flight 965 Accident Investigation Board (1996). Seattle, WA: Boeing.
3 Aeronautica Civil, op. cit., pages 33-34.
4 Tuchman, B. (1981). *Practicing history: Selected essays*. New York: Norton, page 75.

PART II

The New View of Human Error:

Human error is a symptom of trouble deeper inside a system

To explain failure,
do not try to find where people went wrong

Instead, investigate how people's assessments and actions would have made sense at the time, given the circumstances that surrounded them

7. Human Error—the New View

PEOPLE CREATE SAFETY IN COMPLEX SYSTEMS

In the new view of human error:

- Human error is not a cause of failure. Human error is the effect, or symptom, of deeper trouble.
- Human error is not random. It is systematically connected to features of people's tools, tasks and operating environment.
- Human error is not the conclusion of an investigation. It is the starting point.

History is rife with investigations where the label "Human Error" was the conclusion. Paul Fitts, marking the start of aviation human factors in 1947, began to turn this around dramatically. Digging through 460 cases of "pilot error" that had been presented to him, he found that a large part consisted of pilots confusing the flap and gear handles. Typically, a pilot would land and then raise the gear instead of the flaps, causing the airplane to collapse onto the ground and leaving it with considerable damage.

Examining the hardware in the average cockpit, Fitts found that the controls for gear and flaps were often placed next to one another. They looked the same, felt the same. Which one was on which side was not standardized across cockpits. An error trap waiting to happen, in other words. Errors (confusing the two handles) were not incomprehensible or random: they were systematic; connected clearly to features of the cockpit layout.

The years since Fitts (1947) have seen an expansion of this basic idea about human error. The new view now examines not only the engineered hardware that people work with for systemic reasons behind failure, but features of people's operations and organizations as well—features that push people's trade-offs one way or another.

An airline pilot who was fired after refusing to fly during a 1996 ice storm, was awarded 10 million dollars by a jury. The pilot, who had flown for the airline for 10 years, was awarded the money in a lawsuit contending that he had been fired for turning around his turboprop plane in a storm. The pilot said he had made an attempt to fly from Dallas to Houston but returned to the airport because he thought conditions were unsafe.[1]

A hero of the jury (themselves potential passengers probably), they reasoned that this pilot could have decided to press on. But if something had happened to the aircraft as a result of icing, the investigation would probably have returned the finding of "human error", saying that the pilot knowingly continued into severe icing conditions. His trade-off must be understood against the backdrop of a turboprop crash in his company only a few years earlier—icing was blamed in that case.

An example like this confirms that:

- Safety is never the only goal in systems that people operate. Multiple interacting pressures and goals are always at work. There are economic pressures; pressures that have to do with schedules, competition, customer service, public image.
- Trade-offs between safety and other goals often have to made under uncertainty and ambiguity. Goals other than safety are easy to measure (How much fuel will we save? Will we get to our destination?). However, how much people borrow from safety to achieve those goals is very difficult to measure.
- Systems are not basically safe. People in them have to create safety by tying together the patchwork of technologies, adapting under pressure and acting under uncertainty.

Trade-offs between safety and other goals enter, recognizably or not, into thousands of little and larger decisions and considerations that practitioners make every day. Will we depart or won't we? Will we push on or won't we? Will we operate or won't we? Will we go to open surgery or won't we? Will we accept the direct or won't we? Will we accept this display or alarm as indication of trouble or won't we? These trade-offs need to be made under much uncertainty and often under time pressure. This means that:

COMPLEX SYSTEMS ARE NOT BASICALLY SAFE

PEOPLE HAVE TO CREATE SAFETY WHILE NEGOTIATING MULTIPLE SYSTEM GOALS

In the new view on human error:

- People are vital to creating safety. They are the only ones who can negotiate between safety and other pressures in actual operating conditions;
- Human errors do not come unexpectedly. They are the other side of human expertise—the human ability to conduct these negotiations while faced with ambiguous evidence and uncertain outcomes.

INVESTIGATIONS AND THE NEW VIEW ON HUMAN ERROR

In the new view, investigations are driven by one unifying principle:

HUMAN ERRORS ARE SYMPTOMS OF DEEPER TROUBLE

Human error is the starting point of an investigation. The investigation is interested in what the error points to. What are the sources of people's difficulties? Investigations target what lies behind the error—the organizational trade-offs pushed down into individual operating units; the effects of new technology; the complexity buried in the circumstances surrounding human performance; the nature of the mental work that went on in difficult situations; the way in which people coordinated or communicated to get their jobs done; the uncertainty of the evidence around them.

Why are investigations in the new view interested in these things? Because this is where the action is. If we want to learn anything of value about the systems we operate, we must look at human errors as:

- A window on a problem that every practitioner in the system might have;
- A marker in the system's everyday behavior, and an opportunity to learn more about organizational, operational and technological features that create error potential.

Recommendations in the new view:

- Are hardly ever about individual practitioners, because their errors are a symptom of systemic problems that everyone may be vulnerable to;
- Do not rely on tighter procedures because humans need the discretion to deal with complex and dynamic circumstances for which prespecified guidance is badly suited;
- Do not get trapped in promises of new technology. Although it may remove a particular error potential, new technology will likely present new complexities and error traps.
- Try to address the kind of systemic trouble that has its source in organizational decisions, operational conditions or technological features.

PROGRESS ON SAFETY

The new view of human error does not necessarily say that human error does not exist. People screw up the whole time. Goals aren't met; selections are wrongly made; situations get misassessed. In hindsight it is easy to see all of that. People inside the situations themselves may even see that.

But the new view avoids judging people for this. It wants to go beyond saying what people should have noticed or could have done. Instead, the new view seeks to explain "why". The new view of human error wants to understand why people made the assessments or decisions they made—why these assessments or decisions would have made sense from the point of view inside the situation. When you view people's situation from the inside, as much like these people did themselves as you can reconstruct, you may begin to see that they were trying to make the best of their circumstances, under the uncertainty and ambiguity surrounding them. When viewed from inside the situation, their behavior probably made sense—it was systematically connected to features of their tools, tasks and environment. This means that:

> **THE POINT OF AN INVESTIGATION IS NOT TO FIND WHERE PEOPLE WENT WRONG**
>
> **IT IS TO UNDERSTAND WHY THEIR ASSESSMENTS AND ACTIONS MADE SENSE AT THE TIME**

The Field Guide intends to help you investigate human error according to the new view. It intends to help you identify how people's assessments and actions made sense to them, according to their circumstances at the time. In the next chapters, you will find more concrete guidance on how to reconstruct a sequence of events and an unfolding situation the way it looked to the people whose actions and assessments you are now investigating.

Note

1 International Herald Tribune, 15 January 2000.

8. Human Factors Data

Before you can begin to re-assemble the puzzle of human performance, you need data. What data do you need, and where do you get it? The first hunch is to say that you need everything you can get. Human factors, as a field of inquiry, often sits right at the center of an unfolding sequence of events. The people you investigate did not perform in a vacuum—they performed by touching almost every aspect of the system around them.

For example, pilots continued an approach into adverse weather and got into trouble once on the runway. So you need data about weather and how cues about it emerged over time. A nurse administered a ten-fold drug dose to a baby patient who subsequently died. So you need data about drug labeling. You also need data about fatigue, about scheduling pressures, about task saturation and workload, about external distractions, about care-giver-to-patient-ratios relative to time of day or night, about drug administration and double-checking procedures, about physician supervision, and probably much more.

Human factors is not just about humans, just like human error is not just about humans. It is about how features of people's tools and tasks and working environment systematically influence human performance. So you need to gather data about all the features that are relevant to the event at hand. This chapter discusses various sources of human factors data—each with its promises and problems:

- Third-party and historical sources;
- Debriefings of participants themselves;
- Recordings of people's and process performance.

THIRD PARTY AND HISTORICAL SOURCES

A common route that investigators often take is to ask other people about the human performance in question; or to dig into history to find out about their background and past performance. For such data, you can:

- Interview peers or others who can give opinions about the people under investigation;
- Scrutinize training-or other relevant records;
- Document what people did in the days or hours leading up to the mishap.

Finding personal shortcomings

In many investigations, these routes to data are used mainly for a process of "elimination"; as a background check to rule out longer-standing or sudden vulnerabilities that were particular to the people in question. But using third party and historical sources can fuel the bad apple theory: suggesting that the failure is due to personal shortcomings, either temporary or long-running, on part of the people involved in it.

Remember the submarine accident mentioned at the beginning of chapter 2? The admiral testifying about his subordinate commander's performance explained how on a ride a year before he noticed how the commander perhaps did not delegate enough tasks to his crew because of his own great talent. The admiral had had to remind the commander: "let your people catch up".[1]

Hindsight seriously biases the search for evidence about people's personal shortcomings. You now know where people failed, so you know what to look for, and with enough digging you can probably find it too—real or imagined. We are all very good at making sense of a troubling event by letting it be preceded by a troubling history. Much research shows how we construct plausible, linear stories of how failure came about once we know the outcome, which includes making the participants look bad enough to fit the bad outcome they were involved in. Such reactions to failure make after-the-fact data mining of personal shortcomings not just counterproductive (sponsoring the bad apple theory) but probably untrustworthy.

Finding systemic shortcomings

Local shortcomings of individual operators can instead be used as a starting point for probing deeper into the systemic conditions of which their problems are a symptom. Here are some examples:

- From their 72-hour history preceding a mishap, individual operators can be found to have been fatigued. This may not just be a personal problem, but a feature of their operation and scheduling—thus affecting other operators as well.
- Training records may sometimes reveal below average progress or performance by the people who are later caught up in a mishap. But it is only hindsight that connects the two; that enables you to look back from a specific incident and cherry pick associated shortcomings from a historical record. Finding real or imagined evidence is almost pre-ordained because you come looking for it from a backward direction. But this does not prove any specific causal link with actions or assessments in the sequence of events. Training records are a much more interesting source when screened for the things that all operators got trained on, and how and when, as this explains local performance much better. For example, how were they trained to recognize a particular warning that played a role in the mishap sequence? When were they last trained on this? Answers to these questions may reveal more fundamental mismatches between the kind of training people get and the kind of work they have to do.
- Operators may be found to have been overly concerned with, for example, customer satisfaction. In hindsight this tendency can be associated with a mishap sequence: individuals should have zigged (gone around, done it again, diverted, etc.) instead of zagged (pressed on). Colleagues can be interviewed to confirm how customer-oriented these operators were. But rather than branding an individual with a particular bias, such findings point to the entire organization that, in subtle or less subtle ways, has probably been sponsoring the trade-offs that favor other system goals over safety—keeping the practice alive over time.

DEBRIEFINGS OF PARTICIPANTS

What seems like a good idea—ask the people involved in the mishap themselves—also carries a great potential for distortion. This is not because operators necessarily have a desire to bend the truth when asked about their contribution to failure. In fact, experience shows that participants are interested in finding out what went wrong and why, which generally makes them forthright about their actions and assessments. Rather, problems arise because of the inherent features of human memory:

- Human memory does not function like a videotape that can be rewound and played again;
- Human memory is a highly complex, interconnected network of impressions. It quickly becomes impossible to separate actual events and cues that were observed from later inputs;
- Human memory tends to order and structure events more than they were; it makes events and stories more linear and plausible.

Gary Klein has spent many years refining methods of debriefing people after incidents: firefighters, pilots, nurses, and so forth. Insights from these methods are valuable to share with investigators of human error mishaps here.[2]

The aim of a debriefing

Debriefings of mishap participants are intended primarily to help reconstruct the situation that surrounded people at the time and to get their point of view on that situation. Some investigations may have access to a re-play of how the world (for example: cockpit instruments, radar displays, process control panel) looked during the sequence of events, which may seem like a wonderful tool. It must be used with caution, however, in order to avoid memory distortions. Klein proposes the following debriefing order:

1. First have participants tell the story from their point of view, without presenting them with any replays that supposedly "refreshes their memory" but would actually distort it;
2. Then tell the story back to them as investigator. This is an investment in common ground, to check whether you understand the story as the participants understood it;
3. If you had not done so already, identify (together with participants) the critical junctures in the sequence of events (see chapter 9);
4. Progressively probe and rebuild how the world looked to people on the inside of the situation at each juncture. Here it is appropriate to show a re-play (if available) to fill the gaps that may still exist, or to show the difference between data that were available to

people and data that were actually observed by them.

At each juncture in the sequence of events, you want to get to know:

- Which cues were observed (what did he or she notice/see or did not notice what he or she had expected to notice?)
- What knowledge was used to deal with the situation? Did participants have any experience with similar situations that was useful in dealing with this one?
- What expectations did participants have about how things were going to develop, and what options did they think they have to influence the course of events?
- How did other influences (operational or organizational) help determine how they interpreted the situation and how they would act?

Some of Klein's questions to ask

Here are some questions Gary Klein and his researchers typically ask to find out how the situation looked to people on the inside at each of the critical junctures:

Cues	What were you seeing? What were you focusing on? What were you expecting to happen?
Interpretation	If you had to describe the situation to your fellow crewmember at that point, what would you have told?
Errors	What mistakes (for example in interpretation) were likely at this point?
Previous experience/ knowledge	Were you reminded of any previous experience? Did this situation fit a standard scenario? Were you trained to deal with this situation? Were there any rules that applied clearly here? Did you rely on other sources of knowledge to tell you what to do?
Goals	What goals governed your actions at the time? Were there conflicts or trade-offs to make between goals? Was there time pressure?
Taking action	How did you judge you could influence the course of events?

	Did you discuss or mentally imagine a number of options or did you know straight away what to do?
Outcome	Did the outcome fit your expectation?
	Did you have to update your assessment of the situation?

Debriefings need not follow such a tightly scripted set of questions, of course, as the relevance of questions depends on the event under investigation.

Dealing with disagreements and inconsistencies in debriefings

It is not uncommon that operators change their story, even if slightly, when they are debriefed on multiple occasions. Also, different participants who were caught up in the same sequence of events may come with a different take on things. It is difficult to rule out the role of advocacy here—people can be interested in preserving an image of their own contribution to the events that may contradict facts, earlier findings, or statements from others. How should you deal with this as investigator? This mostly depends on the circumstances. But here is some generic guidance:

- Make the disagreements and inconsistencies, if any, explicit in your investigation.
- If later statements from the same people contradict earlier ones, choose which version you want to rely on for your analysis and make explicit why.
- Most importantly, see disagreements and inconsistencies not as impairments of your investigation, but as additional human factors data for it. Mostly, such discontinuities can point you towards goal conflicts that played a role in the sequence of events, and may likely play a role again.

RECORDINGS OF PERFORMANCE DATA

One thing that human error investigations are almost never short of is wishes for more recorded data, and novel ideas and proposals for capturing more performance data. This is especially the case when mishap participants are no longer available for debriefing. Advances in recording what people did have been enormous—there has been a succession of recording materials and strategies, data transfer technologies; everything up to proposals for permanently mounted video cameras in cockpits and other critical workplaces. In aviation, the electronic footprint that a professional pilot leaves during every flight is huge, thanks to monitoring systems now installed in almost every airliner.

Getting these data, however, is only one side of the problem. Our ability to make sense of these data; to reconstruct how people contributed to an unfolding sequence of events, has not kept pace with our growing technical ability to register traces of their behavior. The issue that gets buried easily in people's enthusiasm for new data technologies is that recordings of human behavior—whether through voice (for example Cockpit Voice Recorders) or process parameters (for example Flight Data Recorders)—are never the real or complete behavior.

Recordings represent partial data traces: small, letterbox-sized windows onto assessments and actions that all were part of a larger picture. Human behavior in rich, unfolding settings is much more than the data trace it leaves behind. Data traces point beyond themselves, to a world that was unfolding around the people at the time, to tasks, goals, perceptions, intentions, and thoughts that have since evaporated. The burden is on investigators to combine what people did with what happened around them, but various problems conspire against their ability to do so.

Conventional restrictions

Investigations may be formally restricted in how they can couple recorded data traces to the world (e.g. instrument indications, automation mode settings) that was unfolding around the people who left those traces behind. Conventions and rules on investigations may prescribe how only those data that can be factually established may be analyzed in the search for cause (this is, for example, the case in aviation). Such provisions leave a voice or data recording as only factual, decontextualized and impoverished footprint of human performance.

Lack of automation traces

In many domains this problem is compounded by the fact that today's recordings may not capture important automation-related traces—precisely the data of immediate importance to the problem-solving environment in which many people today carry out their jobs. Much operational human work has shifted from direct control of a process to the management and supervision of a suite of automated systems, and accident sequences frequently start with small problems in human-machine interaction.

Not recording relevant traces at the intersection between people and technology represents a large gap in our ability to understand human contributions to system failure. For example, flight data recorders in many automated airliners do not track which navigation beacons were selected by the pilots, what automation mode control panel selections on airspeed, heading, altitude and vertical speed were made, or what was shown on either of the pilots' moving map displays. This makes it difficult to understand how and why certain lateral or vertical navigational decisions were made, something that can hamper investigations into CFIT accidents (Controlled Flight Into Terrain—an important category of aircraft mishaps).

THE PROBLEM WITH HUMAN FACTORS DATA

One problem with a human error investigation is the seeming lack of data. You may think you need access to certain process or performance parameters to get an understanding not only of what people did, but why. Solutions to this lack may be technically feasible, but socially unpalatable (e.g. video cameras in workplaces), and it actually remains questionable whether these technical solutions would capture data at the right resolution or from the right angles.

This means that to find out about critical process parameters (for instance, what really was shown on that left operator's display?) you will have to rely on interpolation. You must build evidence for the missing parameter from other data traces that you *do* have access to. For example, there may be an utterance by one of the operators that refers to the display ("but it shows that it's to the left..." or something to that effect) which gives you enough clues when combined with other data or knowledge about their tasks and goals.

Recognize that data is not something absolute. There is not a finite amount of data that you could gather about a human error mishap and then think you have it all. Data about human error is infinite, and you will often have to reconstruct certain data from other data, cross-linking and bridging between different sources in order to arrive at what you want to know.

This can take you into some new problems. For example, investigations may need to make a distinction between factual data and analysis. So where is the border between these two if you start to derive or infer certain data from other data? It all depends on what you can factually establish and how factually you establish it. If there is structure behind your inferences—in other words, if you can show what you did and why you concluded what you concluded—it may be quite acceptable to present well-derived data as factual evidence.

Notes

1 International Herald Tribune, 14 March 2001.
2 See: Klein, G. (1998). *Sources of power: How people make decisions.* Cambridge, MA: MIT Press.

9. Reconstruct the Unfolding Mindset

Let us suppose that you now have gathered data that you want to start working with. This chapter will help you close the gap between data and interpretation. What you are about to do—connect data about human performance with interpretations of that performance—is by far the most difficult step in human error analysis. Even science has not figured out exactly how to do this yet. Multiple approaches and methods have been proposed, yet none applicable to the point of perfection, and broad consensus is far off. What we do know is where some of the pitfalls and difficulties lie. Chapter 4 has made you aware of most of them, but here is a brief reminder of what (not) to do:

- Don't jump from data to interpretation in one big step. For example, don't just assert how a few remarks in your data indicate that "people lost situation awareness". No one else will be able to trace why you came to that conclusion. They will just have to take your word for it.
- So leave a trace; a trace that clearly connects your data with your interpretation of it. Other people can then verify your conclusions. This chapter and the next will show you one way how to do this.

FROM CONTEXT TO CONCEPTS

A trace that goes from data to interpretation shows other people how you moved from a context-specific description of what happened to a concept-dependent description. What does that mean?

- **Context-specific** means the data as you have found them. The factual information, as it is sometimes called. For example, upon receiving this or that indication, people threw this or that switch (whatever the indications and switches are in the proper domain language). Context-specific means a minimum of psychological language—no interpretation; no big labels or concepts. You describe what happened in as neutral a way as possible, and you stick with the language of the domain (e.g. aviation, medicine) that people use to describe their own work.

- **Concept-dependent** means you re-inscribe the same events, but then in a different language. This is likely to be the language of human factors. For example, the events above would fit the human factors description of "action slips" because the action in its domain context fits what we know and have conceptualized about action slips so far. Chapter 10 introduces you to a number of such "patterns of failure"—concept-level descriptions of how human performance can go wrong. These descriptions may fit the context of your particular sequence of events.

Five steps to the reconstruction of unfolding mindset

To close the gap between data and interpretation; between context and concepts, you have to take more steps than one. You cannot do it in one assertive jump. And in order for other people to understand your interpretations and learn something of value from your investigation, you would do well to document each of the steps you take. Here they are:

1. The data you have found very likely specifies a sequence of events and activities. Lay out this sequence of events (in context-specific language), using time (and space) as organizing principles.
2. Divide the sequence of events into episodes (still in context-specific language). Each of these episodes may later fit different psychological phenomena, and if you find you did not have the right division intially, you can adjust the boundaries of your various episodes. Human factors concepts are also likely to bind particular episodes together.
3. Find the data you now know to have been available to people during each episode. Of course you will see mismatches between what was available and what people observed or used, but remember that that does not explain anything by itself.
4. Reconstruct people's unfolding mindset: you want to explain why their assessments or actions made sense to them at the time (and forget about emphasizing why they don't make sense to you now—that's not the point). You do this re-establishing people's

knowledge, goals and attention at the time. Use the local rationality principle: people do reasonable things given their knowledge, their objectives, their point of view and limited resources. This is the step that takes most work, and you may need other domain experts to help you out.

5. Link the reconstructed mindset to human factors concepts. If our theory (or model) of the concept in question is mature enough, it should tell you what context-specific data to look for and how to fit it into the concept-dependent description. You may consult chapter 10 or other human factors literature for such possible patterns of failure. Don't get lured by glib, superficial concepts that simply *seem* to fit your data in hindsight. Insist on making clear connections between the concept and the data you have in your hands.

Remember, at all times, what you are trying to do. In order to understand other people's assessments and actions, you must try to attain the perspective of the people who were there at the time. Their decisions were based on what they saw on the inside of the tunnel—not on what you happen to know today:

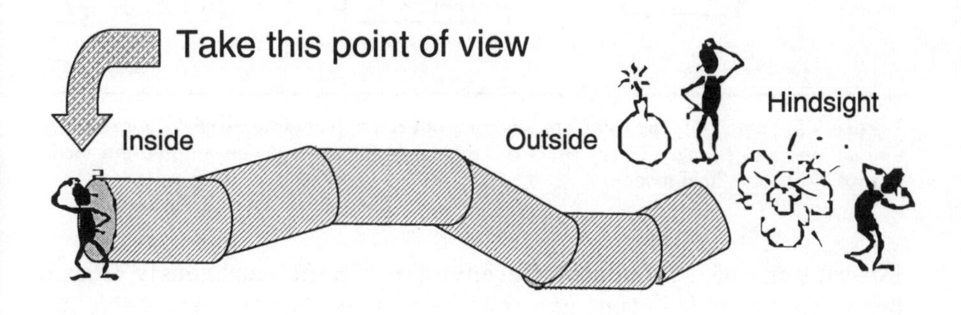

Figure 9.1: See the unfolding world from the point of view of people inside the situation—not from the outside or from hindsight.

1. DESCRIBE THE SEQUENCE OF EVENTS

What do you use to describe the sequence of events? Clearly definable events, such as specific observations and actions by people, or changes in a process that you know happened, may serve as your basic thread:

- People manage situations, giving it direction.
- Situations also develop by themselves.

See figure 9.2 for how this could look—for a shortened, hypothetical sequence of events. You can lay out events in the form of boxes as shown in this figure. This gives you various opportunities to clarify your analysis:

- The distance between boxes can or even should reflect the time between two junctures.
- The way in which the boxes are related to one another vertically (up or down) can also be used as one coding category (for example, a descent into trouble with momentary recoveries).

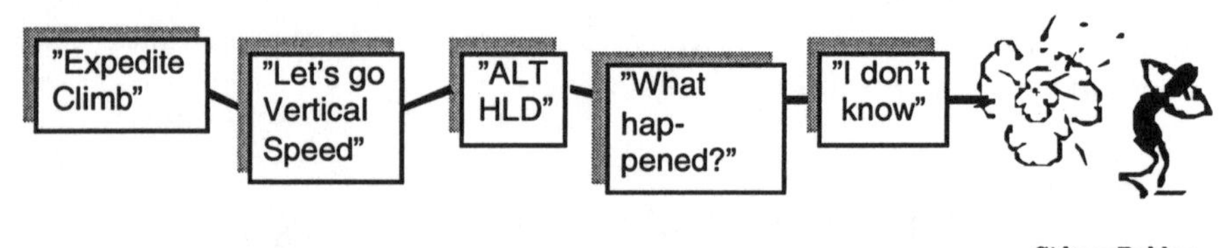

Sidney Dekker

Figure 9.2: Laying out the complete sequence of events, including people's assessments and actions and changes in the process itself (here for example an automation mode change to altitude hold mode).

Descriptions of a sequence of events can benefit enormously from a layout that combines time and space—showing the route to trouble not only over time but also how it meandered through a landscape (the approach path to an airport or harbor, for example). Even there, be sure to coordinate the scales: if actions and assessments are separated by time as well as space, indicate so clearly on your representation of the sequence of events by spacing them apart—to scale.

You may need the distances between various events (in terms of space and/or time) in your subsequent analysis for example to determine task load (how much did people have to accomplish in a certain amount of time) or to find out whether indications in the unfolding landscape were observable from people's vantage point at that moment. A presentation of how a situation unfolded over time (and through space) is the basis for a credible human factors analysis.

How do you identify events in your data?

In order to not get lost in potential masses of raw data about your sequence of events, here are some areas to focus on:

- Events are places, or short stretches of time, where either people or the processes they managed contributed critically to the direction of events and/or the outcome that resulted.
- Events in a sequence of events are places where people did something or (with your knowledge of hindsight) could have done something to influence the direction of events.
- Events are also places where the process did something or could have done something to influence the direction of events—whether as a result of human inputs or not.
- As a rule, what people did and what their processes did is highly interconnected. Finding events in one can or should lead you to events in the other.

Here are some examples of typical events:

- **Decisions** can be obvious events, particularly when they are made in the open and talked about. A point where people chose not to decide anything is still a decision and still an event.
- **Shifts in behavior**. There may be points where people realized that the situation was different from what they believed it to be previously. You can see this either in their remarks or their actions. These shifts are markers where later you want to look for the evidence that people may have used to come to a different realization.
- **Actions to influence the process** may come from people's own intentions. Depending on the kind of data that your domain records or provides, evidence for these actions may not be found in the actions themselves, but in process changes that follow from them. As a clue for a later step, such actions also form a nice little window on people's understanding of the situation at that time.
- **Changes in the process**. Any significant change in the process that people manage must serve as event. Not all changes in a process managed by people actually come from people. In fact, increasing automation in a variety of workplaces has led to the potential for autonomous process changes almost everywhere—for example:

- Automatic shut-down sequences or other interventions;
- Alarms that go off because a parameter crossed a threshold;
- Uncommanded mode changes;
- Autonomous recovery from undesirable states or configurations.

Yet even if they are autonomous, these process changes do not happen in a vacuum. They always point to human behavior around them; behavior that preceded it and behavior that followed it. People may have helped to get the process into a configuration where autonomous changes were triggered. And when changes happen, people notice them or not; people respond to them or not. Such actions, or the lack of them, again give you a strong clue about people's knowledge and current understanding.

The events that were no events

Human decisions, actions and assessments can also be less obvious. For example, people seem to decide, in the face of evidence to the contrary, to not change their course of action; to continue with their plan as it is. With your hindsight, you may see that people had opportunities to recover from their misunderstanding of the situation, but missed the cues, or misinterpreted them.

These "decisions" to continue, these opportunities to revise, may look like clear candidates for events to you. And they are. But they are events only in hindsight. To the people caught up in the sequence of events there was not any compelling reason to re-assess their situation or decide against anything. Or else they would have. They were doing what they were doing because they thought they were right; given their understanding of the situation; their pressures. The challenge for you becomes to understand how this was not an event to the people you were investigating. How their "decision" to continue was nothing more than continuous behavior—reinforced by their current understanding of the situation, confirmed by the cues they were focusing on, and reaffirmed by their expectations of how things would develop.

2. DIVIDE THE SEQUENCE OF EVENTS INTO EPISODES

What is an episode? This is a longer stretch of time that (initially) makes sense from the point of view of the domain. For example, the time taken to taxi out to a runway (or to approach one for landing) is a meaningful chunk of time in which particular actions and assessments need to be made to prepare for the next episode (taking off; landing). To make things easier on yourself, you may want to divide your sequence of events into episodes and consider them more or less separately, at least up to step 4.

When it comes to identifying the beginning of a sequence of events, the issue is often decided implicitly by the availability or lack of evidence.

For example, the beginning of a cockpit voice recording may be where investigative activities start for real, and the end of the recording where they end. Or the beginning is contained in the typical 72-hour or 24-hour histories of what a particular practitioner did and did not do (play tennis, sleep well, wake up early, etc.) before embarking on the fatal journey or operation. Of course even these markers are arbitrary, and the reasons for them are seldom made clear.

There is of course inherent difficulty in deciding what counts as the beginning (especially the beginning—the end of a sequence of events often speaks for itself). Since there is no such thing as a root cause (remember chapter 3), there is technically no such thing as the beginning of a mishap. Yet as investigator you need to start somewhere. Making clear where you start and explaining this choice is the first step toward a structured, well-engineered human error investigation. Here is what you can do:

- Take as the beginning of your first episode the first assessment, decision or action by people close to the mishap, or an event in the process—the one that, according to you, set the sequence of events in motion. This assessment or action can be seen as a trigger for the events that unfold from there.
- Of course the trigger itself has a reason, a background, that extends back beyond the mishap sequence—both in time and in place. The whole point of taking a proximal assessment or action as starting point is not to ignore this background, but to identify concrete points to begin your investigation into them.
- This also allows you to deal with any controversy that may surround your choice of starting point.

Was the pilot's acceptance of a runway change the trigger of trouble? Or was it the air traffic controller's dilemma of having too many aircraft converge on the airport at the same time—something that necessitated the runway change?

Someone can always say that another decision or action preceded the one you marked as your starting point. This is a reminder of what to take into account when analyzing the decision or action you have marked as the beginning. What went on before that? Whatever your choice of beginning, make it explicit. From there you can reach back into history, or over into surrounding circumstances, and find explanations for the decision or action that, according to you, set the sequence of events in motion.

3. FIND WHAT THE WORLD LOOKED LIKE DURING EACH EPISODE

Step 3 is about reconstructing the unfolding world that people inhabited in a straightforward way: find out what their process was doing; what data was available. This is the first step toward coupling behavior and situation—toward putting the observed behavior back into the situation that produced and accompanied it.

Laying out how some of the critical parameters changed over time is nothing new to investigations. Many accident report appendices contain read-outs from data recorders, which show the graphs of known and relevant process parameters. But building these pictures is often where investigations stop today. Tentative references about connections between known parameters and people's assessments and actions are sometimes made, but never in a systematic, or graphic way.

The point of step three is to marry all the events you have identified above with the unfolding process—to begin to see the two in parallel, as an inextricable, causal *dance-à-deux.* The point of step three is to build a picture that shows these connections. How do you do this?

Choosing your datatraces

- Find out how process parameters were changing over time, both as a result of human influences and of the process moving along—make a trace of changing pressures, ratios, settings, quantities, modes, rates, and so forth.
- Find out how the values of these parameters were available to people —dials, displays, knobs that pointed certain ways, sounds, mode annunciations, alarms, warnings. Their availability does not mean people actually observed them: that distinction you will make in step 4.
- Decide which—of all the parameters—counted as a stimulus for the behavior under investigation, and which did not. Which of these indications or parameters, and how they evolved over time, were actually instrumental in influencing the behavior in your mishap sequence?

Here are a few examples from the world most richly endowed with devices for tracking and recording process parameters—commercial aviation: If the outcome of the sequence of events was a stall warning, then airspeed, and what it did over time, becomes a relevant parameter to include. If the outcome involves a departure from the hard surface of a runway, then brake pressure is a parameter to focus on. If the outcome was an automation surprise, then the various mode changes the automation went through, including their annunciations, are what you want to get down.

When are you sure you have covered the parameters you need? After going through the reconstruction of people's unfolding mindsets, you may be left with gaps in your explanation of people's assessments and actions. If so, it may be time to look for more parameters that could have served as critical stimuli to influence people's understanding and behavior—parameters that did not seem obvious before.

Connecting process and behavior

Once you have decided which process parameters to track in their journey towards the outcome, you can use them to embellish the picture you constructed in step 1. With this extended picture, that includes both human assessments and actions and process events, connections can start to emerge between how the world looked and what people did. You have graphically tied the relevant process parameters to the human assessments and actions that evolved in concert with them. This is where one may begin to explain the other—and vice versa. See figure 9.3 for an example.

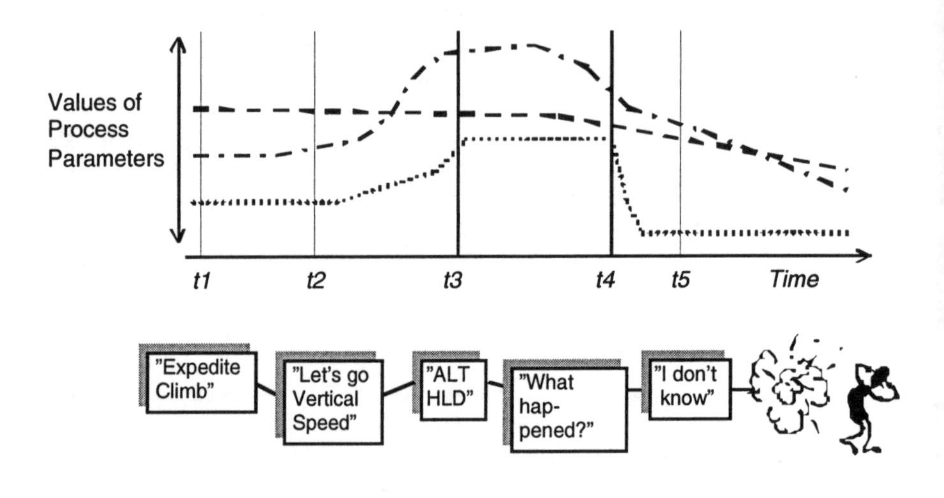

Figure 9.3: Connecting critical process parameters to the sequence of people's assessments and actions and other junctures.

4. IDENTIFY KNOWLEDGE, FOCUS OF ATTENTION AND GOALS

So what, out of all the data available, did people actually see and how did they interpret it? You may have laid out all the possible and relevant parameters, but what did people actually notice? What did they understand their situation to be? The answer lies in three things:

- People have goals. They are in a situation to get a job done; to achieve a particular aim.
- People have knowledge. They use this to interpret what goes on around them.
- People's goals and knowledge together determine their focus of attention. Where people look depends on what they know and what

they want to accomplish.

In this section, we will first look at goals—at conflicts between interacting goals in particular, as those almost always exist in systems operations and have a profound influence on how people decide and act. Then we shift to knowledge and focus of attention.

Goals

Finding what tasks people were working on does not need to be difficult. It often connects directly to how the process was unfolding around them. Setting the navigation systems up for an approach to the airport, for example, is a job that stretches both into what people were saying and doing and to what was happening with the process they managed. Changing a flight plan in the flight management computer is another. To identify what job people were trying to accomplish, ask yourself the following questions:

- What is canonical, or normal at this time in the operation? Jobs relate in systematic ways to stages in a process. You can find these relationships out from your own knowledge or from that of (other) expert operators.
- What was happening in the managed process? Starting from your record of parameters from step 3, you can see how systems were set or inputs were made. These changes obviously connect to the tasks people were carrying out.
- What were other people in the operating environment doing? People who work together on common goals often divide the necessary tasks among them in predictable or complementary ways. There may be standard role divisions, for example between pilot flying and pilot not-flying, that specify the jobs for each. What one operator was doing may give some hints about what the other operator was doing.

If you find that pictures speak more clearly than text, you can create a graphical representation of the major jobs over time, and, if necessary, who was carrying out what. This picture can also give you a good impression of the taskload in the sequence of events. See figure 9.4 for an example. Once you can clearly see what people were busy with, you can begin to understand what they were looking at and why.

It can be difficult to identify the larger goals people were pursuing.

In aviation, you would think an obvious goal is "flight safety". But how do these goals translate to concrete assessments and actions? Sometimes local decisions and actions seem contrary to these goals.

For example, a pilot may do everything to stay visual with an airport where he has just missed an approach. This can lead to all kinds of trouble, for example getting close to terrain, being forced lower by shifting cloud ceilings, getting in conflict with other aircraft, losing bearings, and so forth. So why would anyone do it? In the context in which the pilot was operating, it may actually be an action that lies closest to the goal of flight safety. What kind of country was the airport in? How reliable were the navigation aids around it? How good or understandable were the controllers? How much other traffic was around? How familiar was the pilot with the area? Was there severe turbulence in the clouds? Given this context, the goal of flight safety takes on a different meaning. Achieving flight safety translates to different assessments and actions under different circumstances—ones that may at first seem counterintuitive or counterprocedural.

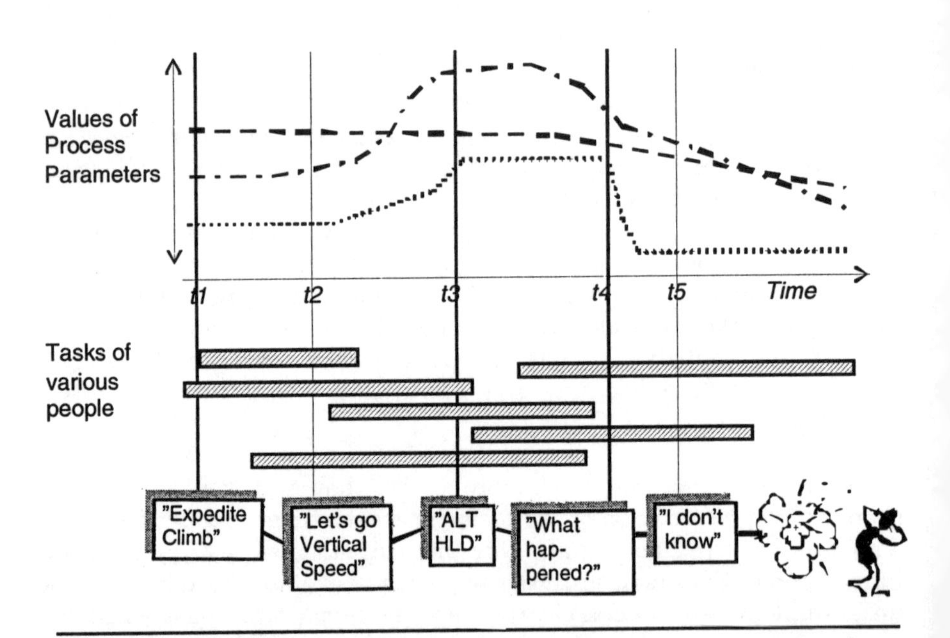

Figure 9.4: Laying out the various (overlapping) tasks that people were accomplishing during the sequence of events.

Follow the goal conflicts

It is seldom the case that only one goal governs what people do. Most complex work is characterized by multiple goals, all of which are active or must be pursued at the same time (on-time performance and safety, for example). Depending on the circumstances, some of these goals may be at odds with one another, producing goal conflicts. Any analysis of human error has to take the potential for goal conflicts into account.

A woman was hospitalized with severe complications of an abdominal infection. A few days earlier, she had seen a physician with complaints of aches, but was sent home with the message to come back in eight days for an ultrasound scan if the problem persisted. In the meantime, her appendix burst, causing infection and requiring major surgery. The woman's physician had been under pressure from her managed care organization, with financial incentives and disincentives, to control the costs of care and avoid unnecessary procedures.[1] The problem is that a physician might not know that a procedure is unnecessary before doing it, or at least doing part of it. Pre-operative evidence may be too ambiguous. Physicians end up in difficult double binds, created by the various organizational pressures.

Although "safety" is almost always cited as an organization's overriding goal, it is never the only goal (and in practice not even a measurably overriding goal), or the organization would have no reason to exist. People who work in these systems have to pursue multiple goals at the same time, which often results in goal conflicts. Goal trade-offs can be generated:

- by he nature of operational work itself
- by the nature of safety and different threats to it
- at the organizational level.

Anesthesiology presents interesting inherent goal conflicts. On the one hand, anesthesiologists want to protect patient safety and avoid being sued for malpractice afterward. This maximizes their need for patient information and pre-operative workup. But hospitals continually have to reduce costs and increase patient turnover, which produces pressure to admit, operate and discharge patients on the same day. Other pressures stem from the need to maintain smooth relationships and working practices with other professionals (surgeons, for example), whose schedules interlock with those of the anesthesiologists.[2]

The complexity of these systems, and of the technology they employ, can also mean that one kind of safety needs to be considered against another. Here is an example of a goal trade-off that results from the nature of safety in different contexts:

The space shuttle Challenger broke up and exploded shortly after lift-off in 1986 because hot gases bypassed O-rings in the booster rockets. The failure has often been blamed on the decision that the booster rockets should be segmented (which created the need for O-rings) rather than seamless "tubes". Segmented rockets were cheaper to produce—an important incentive for an increasingly cash-strapped operation.

The apparent trade-off between cost and safety hides a more complex reality where one kind of safety had to be traded off against another—on the basis of uncertain evidence and unproven technology. The seamless design, for example, could probably not withstand predicted prelaunch bending moments, or the repeated impact of water (which is where the rocket boosters would end up after being jettisoned from a climbing shuttle). Furthermore, the rockets would have to be transported (probably over land) from manufacturer to launch site: individual segments posed significantly less risk along the way than a monolithic structure filled with rocket fuel.[3]

Finally, goal conflicts can be generated by the organizational or social context in which people work. The trade-off between safety and schedule is often mentioned as an example. But other factors produce competition between different goals too, for example:

- Management policies
- Earlier reactions to failure (how has the organization responded to similar situations before?)
- Subtle coercions (to do what the boss wants, not what s/he says)
- Legal liability
- Regulatory guidelines
- Economic considerations (fuel usage, customer satisfaction, public image, and so forth).

Operators can also bring personal or professional interests with them (carreer advancement, avoiding conflicts with other groups), that enter into their negotiations among different goals.

How do you find out about goal conflicts in your investigation? Not all goals are written down in guidance or procedures or job descriptions. In fact, most are probably not. This makes it difficult to trace or prove their contribution to particular assessments or actions. To evaluate the assessments and actions of the people you are investigating, you should:

- List the goals relevant to their behavior at the time (see step 4 in the previous chapter)
- Find out how these goals interact or conflict
- Investigate the factors that influenced people's criterion setting (i.e. what and where was the criterion to pursue the one goal rather than the other, and why was it there?)

Remember the pilot in chapter 7, who refused to fly in severe icing conditions and was subsequently fired? If you were to list the goals relevant in his decision-making, you would find schedule, passenger connections, comfort, airline disruptions and, very importantly: the safety of his aircraft and passengers. In his situation, these goals obviously conflicted. The criterion setting in resolving the goal conflict (by which he decided not to fly) was very likely influenced by the recent crash of a similar aircraft in his airline because of icing. The firing of this pilot sent a message to those who come after him and face the same trade-off. They may decide to fly anyway because of the costs and wrath incurred from chief pilots and schedulers (and passengers even). Yet the lawsuit sent another message, that may once again shift the criterion back a bit toward not flying in such a situation.

It is hard for organizations, especially in highly regulated industries, to admit that these kinds of tricky goal trade-offs arise; even arise frequently. But denying the existence of goal conflicts does not make them disappear. For a human error investigation it is critical to get these goals, and the conflicts they produce, out in the open. If not, organizations easily produce something that looks like a solution to a particular incident, but that in fact makes certain goal conflicts worse.

The possession and application of knowledge

Practitioners usually do not come to their jobs unprepared. They possess a large amount of knowledge in order to manage their operations. The application of knowledge, or using it in context, is not a straightforward activity, however. In order to apply knowledge to manage situations, people need three things:

- Practitioners need to possess the knowledge. Ask yourself whether the right knowledge was there, or whether it was erroneous or incomplete. People may have been trained in ways that leave out important bits and pieces, for example.
- Practitioners need to have the knowledge organized in a way that makes it useable for the situation at hand. People may have learned how to deal with complex systems or complex situations by reading books or manuals about them. This does not guarantee that the knowledge is organized in a way that allows them to apply it effectively in operational circumstances.

The way in which knowledge is organizated in the head seriously affects people's ability to perform well. Knowledge organization is in turn a result of how the material is taught or acquired. Feltovich[4] has investigated how knowledge can be wrongly organized, especially in medicine, leading to misconceptions and misapplications.

One example is that students have learned to see highly interconnected processes as independent from one another, or to treat dynamic processes as static, or to treat multiple processes as the same thing, since that is how they were taught. For example, changes in cardiac output (the rate of blood flow, which is the change of position of volume/minute), are often seen as though they were changes in blood volume. This would lead a student to believe that increases in cardiac output could propagate increases of blood volume, and consequently blood pressure, when in fact increases in blood flow decreases pressure in the veins.

Another example of knowledge organization gone awry happens in Problem-Based Learning (PBL). Although popular in many circles, it carries the risk that students will see one instance of a problem they are confronted with in training as canonical for all instances they will encounter subsequently. This is overgeneralization: treating subtly different problems as similar issues.

- Practitioners also need to activate the relevant knowledge, that is, bring it to bear in context. People can often be shown to possess the knowledge necessary for solving a problem (in a class-room situation, where they are dealing with a textbook problem), but that same knowledge won't "come to mind" when needed in the real world; it remains inert. If material is learned in neat chunks and static ways (books, most computer-based training) but needs to be applied in dynamic situations that call for novel and intricate combinations of those knowledge chunks, then inert knowledge is a risk. In other words, when you suspect inert knowledge, look for mismatches between how knowledge is acquired and how it is (to be) applied.

Training practitioners to work with automation is difficult. Pilots, for example, who learn to fly automated airplanes typically learn how to work the computers, rather than how the computers actually work. They learn the input-output relationships for various well-developed and common scenarios, and will know which buttons to push when these occur on the line. Problems emerge, however, when novel or particularly difficult situations push pilots off the familiar path, when circumstances take them beyond the routine. Knowledge was perhaps once acquired and demonstrated about automation modes or configurations that are seldom used. But being confronted with this in practice means that the pilot may not know what to do—knowledge that is in principle in the head, will remain inert.

Directing attention

What people know and what they try to accomplish jointly determine where they will look; where they will focus their attention. Recognize how this, once again, is the local rationality principle. People are not unlimited cognitive processors (there are no unlimited cognitive processors in the entire universe). People do not know and see everything all the time. So their rationality is bounded. What people do, where they focus, and how they interpret cues makes sense from their point of view; their knowledge, their objectives and their limited resources (e.g. time, processing capacity). Re-establishing people's local rationality will help you understand the gap between data availability (what you discovered in step 3) and what people actually saw or used.

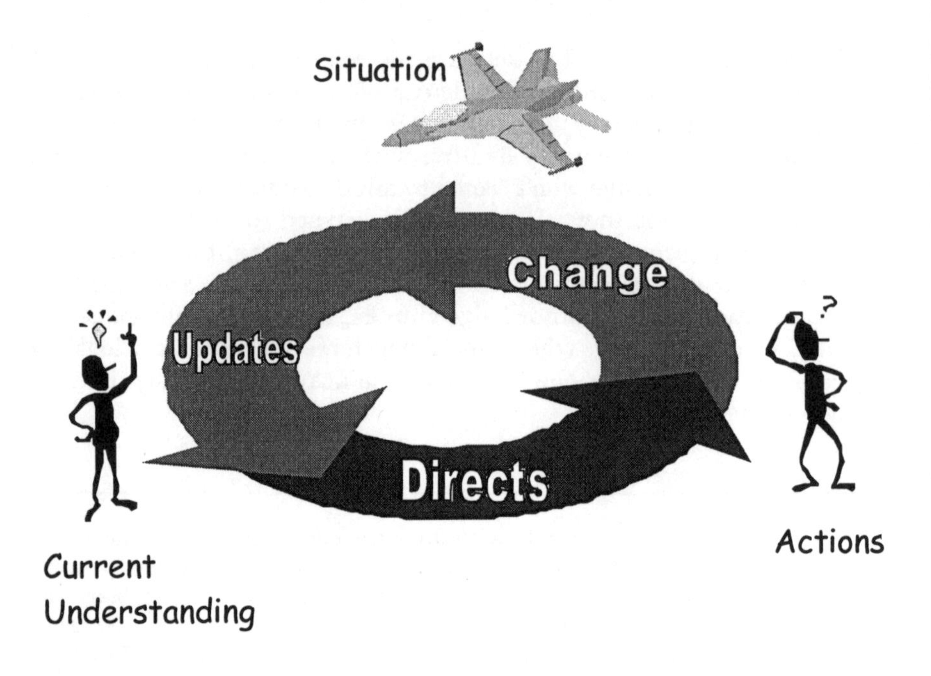

Figure 9.5: We make assessments about the world, updating our current understanding. This directs our actions in the world, which change what the world looks like, which in turn updates our understanding, and so forth (Figure is modeled on Ulrich Neisser's perceptual cycle).

So how does the allocation of attention work? People update their understanding of an unfolding situation on the basis of cues that come in. This understanding in turn directs them to act (or not) in one way or another, which changes the situation (according to expectations or not), which in turn updates people's understanding of what is going on. Figure 9.5 shows this cognitive cycle. The cognitive cycle helps you:

- Reconstruct how people understood their unfolding situation—what they were looking at and how they gave meaning to incoming data and what they were expecting. Ask yourself: what made them focus on certain cues rather than others? What evidence did *they* find to cling onto a hypothesis that you now know was increasingly at odds with the real situation? Yet what in their situation would have made this focus reasonable?

- Look closely at people's actions. Actions are an accurate window on people's current understanding of their situation. What were they trying to find out or influence? What were they driving at?

In dynamic situations, people direct their attention as a joint result of:

- What their current understanding of the situation is, which in turn is determined partly by people's knowledge and goals. Current understanding helps people form expectations about what should happen next (either as a result of their own actions or as a result of changes in the world itself).
- What happens in the world. Particularly salient or intrusive cues will draw attention even if they fall outside people's current interpretation of what is going on.

Keeping up with a dynamic world, in which situations evolve and change, is a demanding part of much operational work. People may fall behind rapidly changing conditions, and update their interpretation of what is happening constantly, trying to follow every little change in the world. Or people become locked in one interpretation, even while evidence around them suggests that the situation has changed. These patterns of failure, vagabonding versus fixation, will be discussed in more detail in the next chapter.

5. STEP UP TO A CONCEPTUAL DESCRIPTION

The last step in your analysis is to build an account of human performance that runs parallel to the one you created in step 1. This time, however, the language that describes the same sequence of events is not one of domain terms, it is one of human factors concepts. When you get to step 5, you may already have formed an idea, even if in non-scientific terms, about what was going on in the minds of people and how this played out in the situation in which they found themselves. Did people seem lost? Did they simply forget something? Were they confused? Was their attention on something that turned out not to be the problem at all? Step 5 intends to formalize these questions into useful answers, by putting everything from the previous steps into concepts that have been developed in human factors so far.

Many investigations encode data only one way: from context-specific to concept-dependent—and often in one big, unverifiable step. For example, a pilot asking a question about where they are headed is taken as evidence of a "loss of situation awareness". But credible, and verifiable, encoding of data in a human factors account of what went on needs to go both ways. In your conceptual account of human performance, you have to link your conclusions back to the context-specifics that, according to your analysis, are an instantiation, an example, of the concept you have converged on. For example, you would need to show that in order to meet the criteria for the concept of "loss of situation awareness" people need to ask questions about direction in your context-specific data. If you cannot, then:

- Either the concept is no good (i.e. it does not make explicit exactly which of your context-specific data match it: it is simply a folk model, see chapter 4)
- Or your data really do not match the concept and you should perhaps take them to another one.

Figure 9.6 shows the steps involved in the reconstruction of unfolding mindset. Starting with the context-specific description of the sequence of events, you work your way up through the analysis, identifying episodes, reconstructing the evolving circumstances around them and finding what people noticed and pursued given their local rationality. From there you step up to the conceptual account of the sequence of events, making sure that you link specific assertions in that description with data or groups of data in your context-specific account.

The shortest possible concept-dependent description may actually serve well as a summary (or, if really necessary "cause" statement) at the end of your investigation. For example, in the case used as illustration in this chapter, the "probable cause statement" could read something like this:

In order to comply with an unusual request from Air Traffic Control, the crew selected an autopilot mode that is not used often. Unfamiliarity with the mode's use in practice combined with poor feedback about mode status in the cockpit. This produced a series of mode changes that went unnoticed, leading to surprising aircraft behavior from which the crew subsequently recovered.

No summary is perfect, and even summaries that double as causal statements are of necessity selective and exclusive. Recognize that any reconstruction of human performance to the level of a concept-dependent account is tentative—and that includes *your* investigation. Later investigation may turn up evidence that calls into question the conclusions that you drew and may motivate the creation of a new or modified account. Do not let this put you off, however. Keep the basic goal of this whole exercise in mind: the target of reconstructing unfolding mindset is to find out why actions and assessments made sense to people at the time.

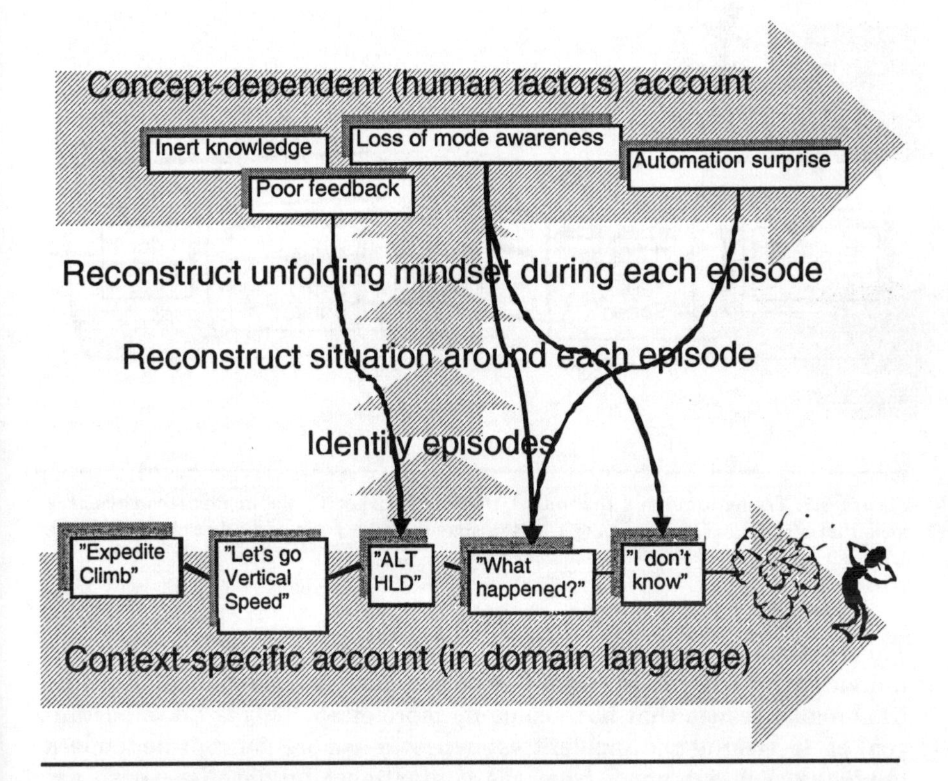

Figure 9.6: Closing the gap from data to interpretation: you must follow and document the various steps between a context-specific account of what happened and a concept-dependent one, linking back the concepts found to specific evidence in the context-specific record.

Around each of the events in the sequence, you have reconstructed what the process looked like. You have been doing what is shown in figure 9.7—covering the tunnel with bits and pieces of data you have found; reconstructing the world as it looked to people on the inside. You have recovered the tasks people were pursuing; the goals they had. All of this may have led you to a better understanding of why people did what they did.

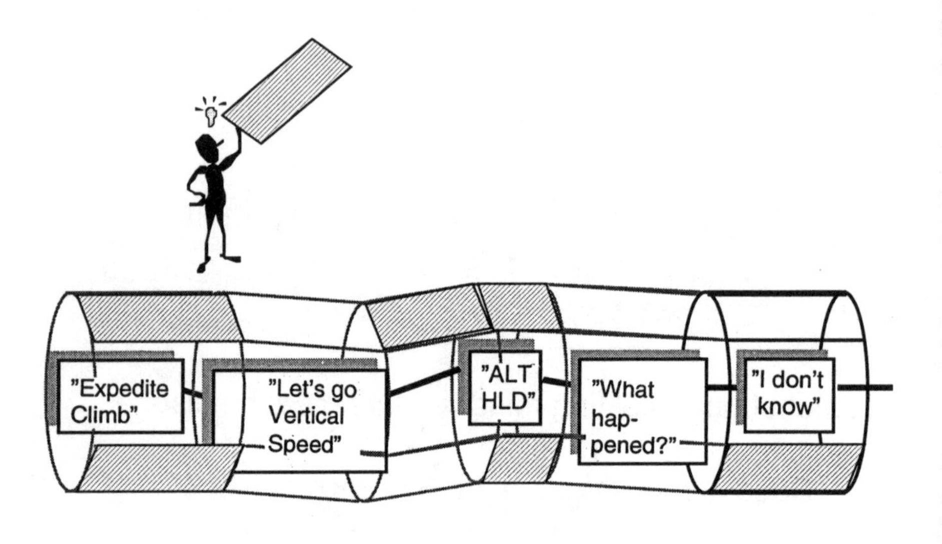

Figure 9.7: Trying to rebuild the tunnel, the way it looked on the inside: reconstructing the situation that surrounded people's assessments and actions and other changes in the process.

The final step, one that perhaps goes beyond the mandate of an individual investigation, would be to see how your sequence of events fits broader issues that have come up more often. This is an important goal of describing the incident sequence in a more concept-dependent language. You get away from the context-specific details—those are stuck in a language that may not communicate well with other context-specific sequences of events. A crucial way to learn from failure is to discover similarities between seemingly disparate events. When people instead stress the differences between sequences of events, learning anything of value beyond the one event becomes difficult.

But in order to see similarities, you have to describe sequences of events in a similar language. That is what the concept-dependent level of description is for. Similarities between accounts of different occurrences can point you to common conditions that helped produce the problem under investigation.

Notes

1 International Herald Tribune, 13 June 2000.
2 See: Woods, D. D., Johanssen, L. J., Cook, R. I., & Sarter, N. B. (1994). Behind human error: Cognitive systems, computers and hindsight. Dayton, OH: CSERIAC, page 63.
3 Vaughan, D. (1996). *The Challenger lauch decision.* Chicago, IL: University of Chicago Press.
4 Feltovich, P. J., Spiro, R. J., & Coulson, R. L. (1993). Learning, teaching, and testing for complex conceptual understanding. In N. Fredericksen, R. Mislevy, & I. Bejar (Eds.). *Test theory for a new generation of tests.* Hillsdale, NJ: Lawrence Erlbaum Associates.

10. Patterns of Failure

The new view of human error does not see human error as the cause of failure. It sees human error as the effect, or symptom, of trouble deeper inside a system. It uses the discovery of human error as the beginning of an investigation, not as its conclusion. The new view seeks to probe events more deeply, change them from a context-specific language into a concept-dependent one through multiple levels of analysis, and finally synthesise across sequences of events to identify patterns of failure.

This chapter is about some of these patterns. These are some of the ways in which good people make honest mistakes; in which people's actions and assessments can go sour; in which the dynamic world can outwit people who work inside of it. The chapter takes you through the following patterns:

- New technology and automation surprises. New technology is supposed to help people do their work better. Sometimes it does the opposite.
- The misconstruction of mindset: A gradual divergence occurs between how people believe their situation to be and how it really is, or has developed.
- Plan continuation: People continue with a plan in the face of cues that, in hindsight, warranted changing the plan.
- Drift into failure. Accidents don't just happen, they are often preceded by an erosion of safety margins that went unnoticed.
- Breach of defenses. A lot needs to go wrong for a system to fall over the edge into breakdown. This model helps you trace defenses that were breached.
- Failures to adapt versus adaptations that fail. Why don't people follow procedures? Following procedures is not just about sticking to rules, it is about context and about substantive cognitive work.
- Stress and workload. Important but often ill-understood terms in complex, dynamic operational worlds. Here is a straightforward way to deal with these phenomena and your potential evidence for them.
- Human-human coordination breakdowns. Where multiple people have to coordinate to get a job done, what are the ways in which their coordination can break down?

Finally, this chapter takes you into your operation's history. History is a good guide, because similar situations may have occurred before, without you knowing it. The chapter concludes with ways to recognize dress rehearsals and contrast cases, both of which can help you better understand the sequence of events currently under investigation.

NEW TECHNOLOGY AND AUTOMATION SURPRISES

Human work in safety-critical domains has almost without exception become work with technology. This means that human-technology interaction is an increasingly dominant source of error. Technology has shaped and influenced the way in which people make errors. It has also affected people's opportunities to detect or recover from the errors they make and thus, in cases, accelerated their journeys towards breakdown.

As is the case with organizational sources of error, human-technology errors are not random. They too are systematically connected to features of the tools that people work with and the tasks they have to carry out. Here is a guide,[1] first to some of the "errors" you may typically find. Then a list of technology features that help produce these errors, and then a list of some of the cognitive consequences of new technology that lie behind the creation of those errors.

More can, and will, be said about technology. The section pays special attention to automation surprises, since these appear to form a common pattern of failure underneath many automation-related mishaps. We will also look at how new technology influences coordination and operational pressures in the workplace.

The new view on the role of technology

How does the new view on human error look at the role of technology? New technology does not remove the potential for human error, but changes it. New technology can give a system and its operators new capabilities, but it inevitably brings new complexities too:

- New technology can lead to an increase in operational demands by allowing the system to be driven faster; harder; longer; more precisely or minutely. Although first introduced as greater protection

against failure (more precise approaches to the runway with a Head-Up-Display, for example), the new technology allows a system to be driven closer to its margins, eroding the safety advantage that was gained.
- New technology is also often ill-adapted to the way in which people do or did their work, or to the actual circumstances in which people have to carry out their work, or to other technologies that were already there.
- New technology often forces practitioners to tailor it in locally pragmatic ways, to make it work in real practice.
- New technology shifts the ways in which systems break down.
- It asks people to acquire more knowledge and skills, to remember new facts.
- It adds new vulnerabilities that did not exist before. It can open new and unprecedented doors to system breakdown.

The new view of human error maintains that:

- People are the only ones who can hold together the patchwork of technologies introduced into their worlds; the only ones who can make it all work in actual practice;
- It is never surprising to find human errors at the heart of system failure because people are at the heart of making these systems work in the first place.

Typical errors with new technology

If people were interacting with computers in the events that led up to the mishap, look for the possibility of the following "errors":

- **Mode error**. The user thought the computer was in one mode, and did the right thing had it been in that mode, yet the computer was actually in another mode.
- **Getting lost** in display architectures. Computers often have only one or a few displays, but a potentially unlimited number of things you can see on them. Thus it may be difficult to find the right page or data set.
- **Not coordinating computer entries**. Where people work together on one (automated) process, they have to invest in common ground by telling one another what they tell the computer, and double-checking each other's work. Under the pressure of circum-

stances or constant meaningless repetition, such coordination may not happen consistently.

- **Overload**. Computers are supposed to off-load people in their work. But often the demand to interact with computers concentrates itself on exactly those times when there is already a lot to do; when other tasks or people are also competing for the operator's attention. You may find that people were very busy programming computers when other things were equally deserving of their attention.
- **Data overload**. People were forced to sort through a large amount of data produced by their computers, and were unable to locate the pieces that would have revealed the true nature of their situation. Computers may also spawn all manner of automated (visual and auditory) warnings which clutter a workspace and proliferate distractions.
- **Not noticing changes**. Despite the enormous visualization opportunities the computer offers, many displays still rely on raw digital values (for showing rates, quantities, modes, ratios, ranges and so forth). It is very difficult to observe changes, trends, events or activities in the underlying process through one digital value clicking up or down. You have to look at it often or continuously, and interpolate and infer what is going on.
- **Automation surprises** are often the end-result: the system did something that the user had not expected. Especially in high tempo, high workload scenarios, where modes change without direct user commands and computer activities are hard to observe, people may be surprised by what the automation did or did not do.

Computer features

What are some of the features of today's technology that contribute systematically to the kinds of errors discussed above?

- Computers can make things "invisible"; they can hide interesting changes and events, or system anomalies. The presentation of digital values for critical process parameters contributes to this "invisibility". The practice of showing only system *status* (what mode it is in) instead of *behavior* (what the system is actually doing; where it is going) is another reason. The interfaces may look simple or appealing, but they can hide a lot of complexity.
- Computers, because they only have one or a few interfaces (this is

called the "keyhole problem"), can force people to dig through a series of display pages to look for, and integrate, data that really are required for the task in parallel. A lot of displays is not the answer to this problem of course, because then navigation across displays becomes an issue. Rather, each computer page should present aids for navigation (How did I get here? How do I get back? What is the related page and how do I get there?). If not, input or retrieval sequences may seem arbitrary, and people will get lost.

- Computers can force people into managing the interface (How do I get to that page? How do we get it into this mode?) instead of managing the safety-critical process (something the computer was promised to help them do). These extra interface management burdens often occur during periods of high workload.
- Computers can change mode autonomously or in other ways that are not commanded by the user (these mode changes can for example result from pre-programmed logic, much earlier inputs, inputs from other people or parts of the system, and so forth).
- Computers ask people typically in the most rudimentary or syntactic ways to verify their entries (Are you sure you want to go to X? We'll go to X then) without addressing the meaning of their request and whether it makes sense given the situation. And when people tell computers to proceed, it may be difficult to make them stop. All this limits people's ability to detect and recover from their own errors.
- Computers are smart, but not that smart. Computers and automation can do a lot for people—they can almost autonomously run a safety-critical process. Yet computers typically know little about the changing situation around them. Computers assume a largely stable world where they can proceed with their pre-programmed routines even if inappropriate; they dutifully execute user commands that make no sense given the situation; they can interrupt people's other activities without knowing they are bothersome.

Cognitive consequences of computerization

The characteristics of computer technology discussed above shape the way in which people assess, think, decide, act and coordinate, which in turn determines the reasons for their "errors":

- Computers increase demands on people's memory (What was this mode again? How do we get to that page?).

- Computers ask people to add to their package of skills and knowledge for managing their processes (How to program, how to monitor, and so forth). Training may prove no match to these new skill and knowledge requirements: much of the knowledge gained in formal training may remain inert (in the head, not practically available) when operators get confronted with the kinds of complex situations that call for its application.
- Computers can complicate situation assessment (they may show digital values or letter codes instead of system behavior) and undermine people's attention management (how you know where to look when).
- By new ways of representing data, computers can disrupt people's traditionally efficient and robust scanning patterns;
- Through the limited visibility of changes and events, the clutter of alarms and indications, extra interface management tasks and new memory burdens, computers increase the risk of people falling behind in high tempo operations.
- Computers can increase system reliability to a point where mechanical failures are rare (as compared with older technologies). This gives people little opportunity for practicing and maintaining the skills for which they are, after all, partly still there: managing system anomalies.
- Computers can undermine people's formation of accurate mental models of how the system and underlying process work, because working the safety-critical process through computers only exposes them to a superficial and limited array of experiences.
- Computers can mislead people into thinking that they know more about the system than they really do, precisely because the full functionality is hardly ever shown to them (either in training or in practice). This is called the knowledge calibration problem.
- Computers can force people to think up strategies (programming "tricks") that are necessary to get the task done. These tricks may work well in common circumstances, but can introduce new vulnerabilities and openings to system breakdown in others.

New technology and operational pressures

Are new technology and operational pressures related to one another? The answer is yes. The introduction of new technology can increase the operational requirements and expectations that organizations impose on people. Organizations that invest in new technologies often unknowingly exploit the advances by requiring operational personnel to do more, do it more quickly, do it in more complex ways, do it with fewer other resources, or do it under less favorable conditions.

Larry Hirschorn talks about a law of systems development, which is that every system always operates at its capacity. Improvements in the form of new technology get stretched in some way, pushing operators back to the edge of the operational envelope from which the technological innovation was supposed to buffer them.

In operation Desert Storm, during the Gulf War, much of the equipment employed was designed to ease the burden on the operator, reduce fatigue, and simplify the tasks involved in combat. Instead these advances were used to demand more from the operator. Almost without exception, technology did not meet the goal of unencumbering the military personnel operating the equipment. Weapon and support systems often required exceptional human expertise, commitment and endurance. The Gulf War showed that there is a natural synergy between tactics, technology and human factors: effective leaders will exploit every new advance to the limit.[2]

Automation surprises

Automation surprises are cases where people thought they told the automation to do one thing, while it is actually doing another. For example, the user dials in a flight path angle to make the aircraft reach the runway, whereas the automation is actually in vertical speed mode—interpreting the instruction as a much steeper rate of descent command rather than a flight angle. Automation may be doing something else because of many reasons, among them:

- It is in a different mode from what people expected or assumed when they provided their instructions.
- It shifted to another mode after having received instructions.
- Another human operator has overriden the instructions given earlier.

Automation surprises appear to occur primarily when the following circumstances are present:

- Automated systems act on their own, that is, without immediately preceding user input. The input may have been provided by someone else, or a while back, or the change may be the result of pre-programmed system logic.
- There is little feedback about the behavior of the automated system that would help the user discover the discrepancy. Feedback is mostly status-based, telling the user—in criptic abbreviations—what state the system is in ("I am now in V/S"), not what it is actually doing or what it will do in the near future. So even though these systems behave over time, they do not tell the user of their behavior, only about their status.
- Event-driven circumstances, that is, where the situation unfolding around people governs how fast they need to think, decide, act, often help with the creation of automation surprises. Novel situations, ones that people have not encountered before, are also likely to help produce automation surprises.
- Intervening in the behavior after an automation surprise may also be hard. In addition to being "silent" (not very clear about their behavior), automation is often hard to direct: it is unclear what must be typed or tweaked to make it do what the user wants. People's attention shifts from managing the process to managing the automation interface.

People generally have a hard time discovering that the automation is not behaving according to their intentions. There is little evidence that people are able to pick the mismatch up from the displays or indications that are available—again because they are often status-oriented and tiny. People will discover that the automation has been doing something different when they first notice strange or unexpected process behavior. This is when the point of surprise has actually been reached, and it may take quite a while—34 hours in one case (see the next section). During many circumstances, however, especially in aviation or critical care medicine or many forms of process control industries, people do not have so much time to discover the discrepancy. Serious consequences may have already ensued by that time.

Automation and coordination

In efforts to sum up the issues with automation, we often refer to people slipping "out-of-the-loop". The problem is thought to be that "as pilots perform duties as system monitors, they will be lulled into complacency, lose situational awareness, and not be prepared to react in a timely manner when the system fails".

There is, however, little evidence that this happens, and no evidence at all that this leads to the serious accidents. First, automation hardly ever "fails" in the one-zero sense. In fact, manufacturers consistently point out in the wake of accidents how their automation behaved as designed. An out-of-the-loop problem—in the sense that people are unable to intervene effectively "when the system fails" after a long period of only monitoring—does not lie behind the problems that occur. In fact, the opposite appears true.

Bainbridge wrote about the ironies of automation in 1987. She observed that automation took away the easy parts of a job, and made the difficult parts more difficult. Automation counted on human monitoring, but people are bad at monitoring for very infrequent events. Indeed, automation did not fail often, which limited people's ability to practice the kinds of breakdown scenarios that still justified their presence in the system. The human is painted as a passive monitor, whose greatest safety risks would lie in deskilling, complacency, vigilance decrements and the inability to intervene successfully in deteriorating circumstances.

These problems occur, obviously, but these are not the behaviors that precede accidents we have seen happen with, for example, automated airliners over the past two decades. Instead, pilots have roles as active supervisors, or managers, who need to coordinate a suite of human and automated resources in order to get an aeroplane to fly. Yet pilots' ability to coordinate their activities with those of computers and other pilots on the flight deck is made difficult by silent and strong (or powerful and independent) automation; by the fact that each human has private access to the automation (each pilot has his/her own flight management system control/display unit); and because demands to coordinate with the automation accrue during busy times when a lot of communication and coordination with other human crewmembers is also needed.

The same features, however, that make coordination difficult make it critically necessary. This is where the irony lies. Coordination is necessary to invest in a shared understanding of what the automated system has been told to do (yet difficult because it can be told to do things separately by any pilot and then go on its way without showing much of what it is doing). Coordination is also necessary to distribute work during busier, higher pressure operational episodes, but such delegation is difficult because automation is hard to direct and can shift a pilot's attention from flying the aircraft to managing the interface.

Accidents are preceded by practitioners being active managers—typing, searching, programming, planning, responding, communicating, questioning—trying to coordinate their intentions and activities with those of other people and the aircraft automation, exactly like they would in the pursuit of success and safety. What consistently seems to lie behind these mishaps is a coordination breakdown between human and automated cockpit crewmembers. A breakdown in teamplay between humans and machines leads to a series of miscommunications and misassessments; a string of commissions and omissions. It can also lead to the misconstruction of mindset, where people think they have told the automation one thing, while it actually is doing another, taking them places or to corners of the operating envelope they had never intended to visit.

THE MISCONSTRUCTION OF MINDSET

One June 10, 1995, a passenger ship named *Royal Majesty* left St. Georges in Bermuda. On board were 1509 passengers and crewmembers who had Boston as destination—677 miles away, of which more than 500 would be over open ocean. Innovations in technology have led to the use of advanced automated systems on modern maritime vessels. Shortly after departure, the ship's navigator set the ship's autopilot in the navigation (NAV) mode. In this mode, the autopilot automatically corrects for the effects of set and drift caused by the sea, wind and current in order to keep the vessel within a preset distance of its programmed track. Not long after departure, when the *Royal Majesty* dropped off the St. Georges harbor pilot, the navigator compared the position data displayed by the GPS (satellite-based) and the Loran (ground/radio-based) positioning systems. He found that the two sets of data indicated positions within about a mile of each other—the expected accuracy in that part of the world. From there on, the Royal Majesty followed its programmed track (336 degrees), as indicated on the automatic radar plotting aid. The navigator plotted hourly fixes on charts of the area using position data from the GPS. Loran was used only as a back-up system, and when checked early on, it revealed positions about 1 mile southeast of the GPS position.

About 34 hours after departure, the *Royal Majesty* ran aground near Nantucket Island. It was about 17 miles off course. The investigation found that the cable leading from the GPS receiver to its antenna had come loose and that the GPS unit (the sole source of navigation input to the autopilot) had defaulted to dead-reckoning (DR) mode about half an hour after departure. Evidence about the loss of signal and default to DR mode was minimal, contained in a few short beeps and a small mode annunciation on a tiny LCD display meters from where the crew normally worked. In DR mode, there was no more correction for drift. A northeasterly wind had blown the *Royal Majesty* further and further west.

People generally interpret cues about the world on the basis of what they have told their automated systems to do, rather than on the basis of what their automated systems are actually doing. In fact, people do not act on the basis of reality, they act on the basis of their perception of reality. Once they have programmed their ship to steer to Boston in NAV mode, they may interpret cues about the world as if the ship is doing just that. Evidence about a mismatch has to be very compelling for people to break out of the misconstruction of mindset. They have no expectation of a mismatch (the system has behaved reliably in the past), and such feedback as there is (a tiny mode annunciation) is not compelling when viewed from inside the situation.

A label that has become popular for these types of situations—loss of situation awareness—is also problematic. The traditional idea is that we process information from the world around us and form a picture of what is going on on the basis of it. Such information processing is typically thought to go through several stages (for example perceiving elements in the situation, processing their meaning and understanding their future implications) before arriving at full situation awareness. (see figure 10.1).

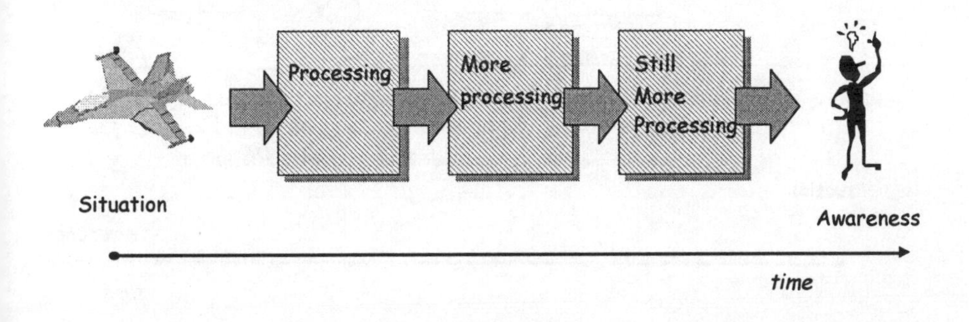

Figure 10.1: The traditional notion of situation awareness: we process information from the world until we arrive at awareness, or a mental picture of what is going on.

A "loss of situation awareness" may occur when our information processing is hampered in some way, for example by high stress or workload. (see figure 10.2). There are major problems with this notion:

- It portrays people as passive recipients of whatever the world throws at them, and everything is OK as long as their mental processing can keep up.
- In this model people make no active contribution to their understanding of the world, and no active contribution to changing the world itself—which they certainly do in reality. For example, they move around in the world; change and tweak things to make it reveal more about itself; influence it to make it slow down; they decide to look in some places rather than others.
- People do not perceive elements in a situation and only then set out to make sense of them by gradually adding meaning along an intra-psychic information highway. If people would perceive individual "elements", they would get pummeled by the world. In reality people perceive patterns, structures. People give meaning to the world simply by looking at it. People rely on their experience to recognize key patterns that indicate what is happening; where things are going.

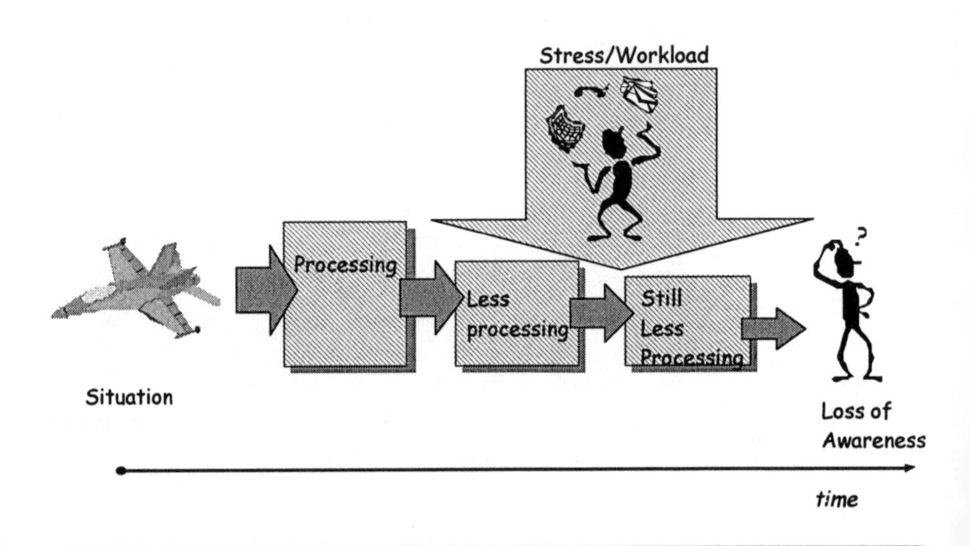

Figure 10.2: In the traditional notion, a loss of situation awareness is presumed to occur through pressures and difficulties in processing information.

Finally, we cannot "lose awareness" other than by becoming physically unconscious. There is no such thing as a mental vacuum. We always have *some* idea of where we are; of what our system and process are doing. We cannot help but give meaning to incoming cues, based on what we already know or expect to happen.

IF YOU LOSE SITUATION AWARENESS, WHAT REPLACES IT?

THERE IS NO SUCH THING AS A MENTAL VACUUM

Indeed, the question "what is happening now?" has such an idea behind it: people had expectations of what was going to happen. We could not live without constantly building and modifying our idea of the changing world around us, influencing our situation on the basis of it, and then receiving new information which updates that understanding once again. When, in hindsight, you uncover a mismatch between how people understood their situation to be, and how you now know it really was, realize that nothing was lost—certainly not the awarenss of the people you are investigating. You must not point out how people at another time and place did not know what you know today (calling it *their* "loss of situation awareness"), because it explains nothing.

PLAN CONTINUATION

Plan continuation is the tendency to continue with an original plan of action in the face of cues that together, in hindsight, warranted changing the plan. In cases of plan continuation, people's understanding of the situation gradually diverges from the situation as it actually turned out to be. This happens almost always because:

- Early and sustained cues that suggest the plan is safe, are compelling and unambiguous;
- Later cues that suggest the situation is changing are much weaker, difficult to process, ambiguous or contradictory.

Plan continuation is a highly typical pattern that is a result of people's normal interactions with a dynamic world, where indications of what is going on emerge over time, interact, and may be of differing quality, persuasiveness, or ease of interpretation.

The pattern is typical because people in dynamic worlds always face a trade-off between changing their assessments and actions with every little change (or possible indication of change) in the world, versus providing some stability in interpretation to better manage and oversee an unfolding situation; creating a framework in which to place newly incoming information. There are errors of judgment on both ends. On the one, people run out of time, they fall behind the world. On the other, people can get fixated, they do not revise their assessment in the face of cues that (in hindsight) suggested it would be good to do so.

Earlier, fixation or plan continuation was often called "confirmation bias". This, however, is not a very helpful label to understand human performance in these situations. First, the confirmation bias says that people actively and exclusively seek information that confirms their hypothesis about what is going on around them. There is little data from complex, dynamic worlds that people seek only that, or seek predominantly actively at all. Instead, people interpret incoming cues on the basis of their current understanding of the situation (see the cognitive cycle). Second, the confirmation bias would suggest that people do not even consider alternative hypotheses about what is going on. This is seldom true. People almost always have alternative hypotheses or options at the ready (Can we land? What if we can't because of the weather, what do we do then?).

DRIFT INTO FAILURE

Confronted with failure, it can be easy to see people's behavior as deficient, as unmotivated, as not living up to what you would expect from operators in their position. "Complacency" and "negligence" are popular terms here. Over time, you think that people seem to have lost respect for the seriousness of their jobs—they start reading newspapers while driving their trains or flying their aircraft, they do not double-check before beginning an amputation.

Departures from the routine that become routine

Figure 10.3 shows what really may be going on here and why complacency or negligence is not only a judgment, but also an incomplete label. Departures from some norm or routine may at any one moment seem to occur because people are not motivated to do otherwise. But theirs is often not the first departure from the routine. Departures from the routine that have become routine can include anything from superficial checklist reading, to cutting other corners to save time, to signing off equipment or people without all official criteria met.

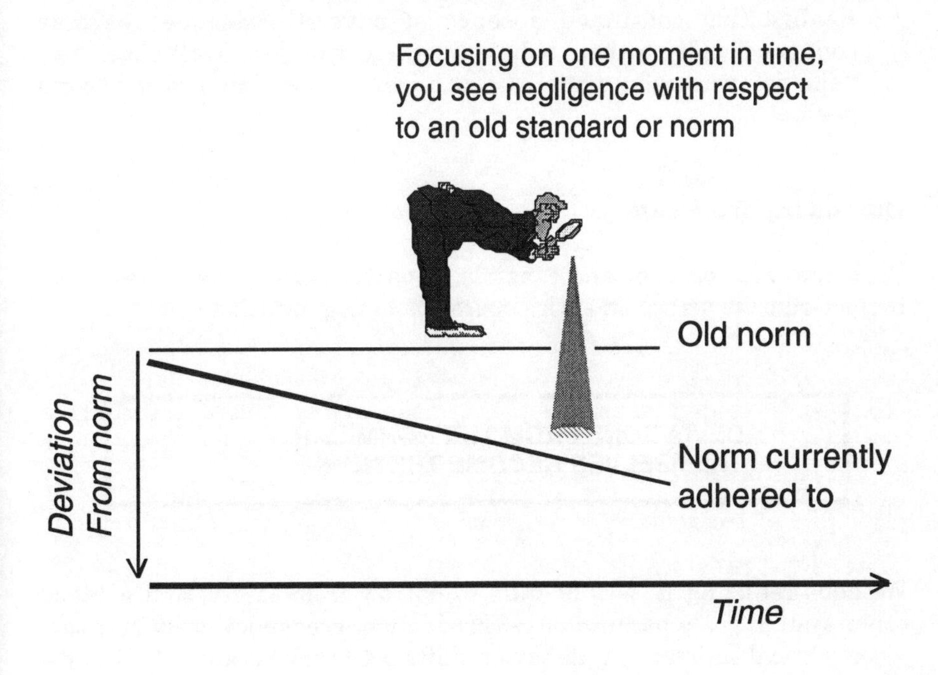

Figure 10.3: At a particular moment in time, behavior that does not live up to some standard may look like complacency or negligence. But deviance may have become the new norm across an entire operation or organization.

When you suspect "complacency", what should you do?

- Recognize that it is often compliance that explains people's behavior: compliance with norms that evolved over time—not deviance. What people were doing was reasonable in the eyes of those on the inside the situation, and others doing the same work every day.

- Find out what organizational history or pressures exist behind these routine departures from the routine; what other goals help shape the new norms for what is acceptable riska and behavior.
- Understand that the rewards of departures from the routine are probably immediate and tangible: happy customers, happy bosses, money made, and so forth. The potential risks (how much did people borrow from safety to achieve those goals?) are unclear, unquantifiable or even unknown.
- Realize that continued absence of adverse consequences may confirm people in their beliefs (in their eyes justified!) that their behavior was safe, while also achieving other important system goals.

Borrowing from safety

With rewards constant and tangible, departures from the routine may become routine across an entire operation or organization.

DEVIATIONS FROM THE NORM CAN THEMSELVES BECOME THE NORM

Without realizing it, people start to borrow from safety, and achieve other system goals because of it—production, economics, customer service, political satisfaction. Behavior shifts over time because other parts of the system send messages, in subtle ways or not, about the importance of these goals. In fact, organizations reward or punish operational people in daily trade-offs ("We are an ON-TIME operation!"), focusing them on goals other than safety. The lack of adverse consequences with each trade-off that bends to goals other than safety, strengthens people's tacit belief that it is safe to borrow from safety.

In "The Challenger Launch Decision", Diane Vaughan has carefully documented how an entire organization started borrowing from safety—reinforced by one successful Space Shuttle Launch after the other, even if O-rings in the solid rocket boosters showed signs of heat damage. The evidence for this O-ring "blow-by" was each time looked at critically, assessed against known criteria, and then decided upon as "acceptable". Vaughan has called this repeated process "the normalization of deviance": what was deviant earlier, now became the new norm. This was thought to be safe: after all, there were two O-rings: the system was redundant. And if past launches were anything to go by (the most tangible evidence for success), future safety would be guaranteed. The Challenger Space Shuttle, launched in cold temperatures in January 1986, showed just how much NASA had been borrowing from safety: it broke up and exploded after lift-off because of O-ring blow-by.[3]

The problem with complex, dynamic worlds is that safety is not a constant. Past success while departing from a routine is not a guarantee for future safety. In other words, a safe outcome today is not a guarantee of a safe outcome tomorrow, even if behavior is the same. Circumstances change, and so do the safety treats associated with them. Doing what you do today (which could go wrong but did not) does not mean you will get away with it tomorrow. The dynamic safety threat is pictured in the figure below.

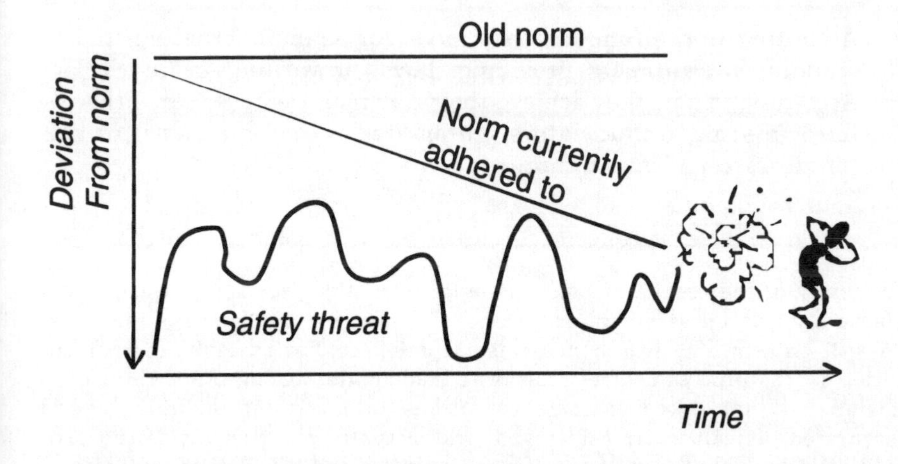

Figure10.4: Murphy's law is wrong.[4] What can go wrong usually goes right, and over time we come to think that a safety threat does not exist or is not as bad. Yet while we adjust our behavior to accommodate other system pressures (e.g. on-time performance), safety threats vary underneath, setting us up for problems sometime down the line.

MURPHY'S LAW IS WRONG

WHAT CAN GO WRONG USUALLY GOES RIGHT, BUT THEN WE DRAW THE WRONG CONCLUSION

Of course, in order to be wrong, Murphy's law has to be right—in the end, that is. That which can go wrong must at some point go wrong, otherwise there would not even be a safety threat.

DEFENSES BREACHED

So when *do* things finally go wrong? One way to think about the breakthrough into failure is in the form of the breaching or by-passing of defenses. As explained in chapter 3, safety-critical organizations invest heavily in multiple layers of defense against known or possible failure trajectories. If failures do happen, then something has to be wrong with these layers of defense. Defenses can be thought of in many ways:

- According to the function they have, for example creating understanding or awareness, providing alarms or warnings or interlocks;
- According to how they achieve their function. Defenses can either be hard (alarms, annunciators, automated reconfigurations) or soft (briefings, certification, training).

The story of the escape of huge amounts of methyl isocyanate (MIC) from Union Carbide's pesticide plant in Bhopal, India, in 1984 is one of many by-passed, broken, breached or non-existent defenses. For example, instrumentation in the process control room was inadequate: among other things, its design had not taken extreme conditions into account: meters pegged (saturated) at values far below what was actually going on inside the MIC tank. Defenses that could have stopped or mitigated the further evolution of events either did not exist or came up short. For example, none of the plant operators had ever taken any emergency procedures training. The tank refrigeration system had been shut down and was now devoid of liquid coolant; the vent gas scrubber was designed to neutralize escaping MIC gasses of quantities 200 times less and at lower temperatures than what was

actually escaping; the flare tower (that would burn off escaping gas and was itself intact) had been disconnected from the MIC tanks because maintenance workers had removed a corroded pipe and never replaced it. Finally, a water curtain to contain the gas cloud could reach only 40 feet up into the air, while the MIC billowed from a hole more than 100 feet up.

An analogy used to illustrate the breaching of defenses is that of "swiss cheese": several layers of defense that are not completely solid but can each allow failures to pass through:[5] As long as the holes do not all line up, failures can be halted somewhere before producing an accident. (figure 10.5.).

Figure 10.5: The "swiss cheese" analogy of defenses that are breached (after Reason, 1997). The analogy itself does not explain what the holes consist of or why they line up to let a failure become an accident.

The layers of defense are not static or constant, and not independent of each other either. They can interact, support or erode one another. The swiss cheese analogy is useful to think about the complexity of failure, and, conversely, about the effort it takes to make and keep a system safe. It can also help structure your search for distal contributors to the mishap. But the analogy itself does not explain:

- where the holes are, or what they consist of.
- why the holes are there in the first place.
- why the holes change over time, both in size and location.
- how the holes get to line up to produce an accident.

This is up to you, as investigator, to find out for your situation. Also, investigating which layers of defense were breached or by-passed reveals more than just the reasons for a particular failure. The existence of defenses (or the holes you find in them) carries valuable information about the organization's current beliefs, and the nature of its understanding about vulnerabilities that threaten safety. This can open up opportunities for more fundamental countermeasures.

FAILURES TO ADAPT AND ADAPTATIONS THAT FAIL

Why don't people follow the rules? Systems would be so much safer if they did. This is often our naive belief. Procedure-following equals safety. The reality is not quite so simple. In fact, if it were, here is an interesting thought experiment.

A landing checklist has all the things on it that need to be done to get an aircraft ready for landing. Pilots read the items off it, and then make them happen in the systems around them. For example:

- Hydraulic pumps ON
- Altimeters SET
- Flight Instruments SET
- Seat Belt sign ON
- Tail de-ice AS REQ.
- Gear DOWN
- Spoilers ARMED
- Auto brakes SET
- Flaps and slats SET

There is no technical reason why computers could not accomplish these items today. Automating a before-landing checklist is, as far as software programming goes, entirely feasible. Why do we rely on unreliable people to accomplish these items for us? The answer is context. Not every approach is the same. The world is complex and dynamic. Which items come when, or which ones can or must be accomplished, perhaps ahead of others, is something that is determined in part by the situation in which people find themselves. Computers are not sufficiently sensitive to the many subtle variations in context. They would go through that list rigidly and uncompromisingly (precisely the way you may think people should have acted when you sort through the rubble of their mishap), ignoring features of the situation that make accomplishing the item at that time entirely inappropriate. Applying procedures is not simple rule-following—neither by computers nor by people. Applying procedures is a substantive cognitive activity.

Think for example of an inflight fire or other serious malfunction where pilots must negotiate between landing overweight or dumping fuel (two things you simply can't do at the same time), while sorting through procedures that aim to locate the source of trouble—in other words, doing what the book and training and professional discipline tells them to do. If the fire or malfunction catches up with the pilots while they are still airborne, you may say that they should have landed instead of bothered with anything else. But it is only hindsight that allows you to say that.

Situations may (and often do) occur where multiple procedures need to be applied at once, because multiple things are happening at once. But items in these various procedures may contradict one another. There was one case, for example, where a flight crew noticed both smoke and vibrations in their cockpit. There was no procedure that told them how to deal with the combination of symptoms. Adaptation, improvisation was necessary to deal with the situation.

There may also be situations, like the one described above, where hindsight tells us that the crew should have adapted the procedure, shortcut it, abandon it. But they failed to adapt; they stuck to the rules rigidly. So the fire caught up with them. There are also standard situations where rigid adherence to procedures leads to less safety. The example from chapter three, explaining the error trap of not arming the ground spoilers on an aircraft, shows that people actually create safety by adapting procedures in a locally pragmatic way.

The pilot who adapted successfully was the one who, after years of experience on a particular aircraft type, figured out that he could safely arm the spoilers 4 seconds after "gear down" was selected, since the critical time for potential gear compression was over by then. He had refined a practice whereby his hand would go from the gear lever to the spoiler handle slowly enough to cover 4 seconds—but it would always travel there first. He then had bought himself enough time to devote to subsequent tasks such as selecting landing flaps and capturing the glide slope. This obviously "violates" the original procedure, but the "violation" is actually an investment in safety, the creation of a strategy that help forestall failure.

Applying procedures can thus be a delicate balance between:

- Adapting procedures in the face of either unanticipated complexity or a vulnerability to making other errors. But people may not be entirely successful at adapting the procedure, at least not all of the time. They will then be blamed for not following procedures; for improvising.
- Sticking to procedures rigidly, and discovering that adapting them would perhaps have been better. People will then be blamed for not being flexible in the face of a need to be so.

Notice the double bind that practitioners are in here: whether they adapt procedures or stick with them, with hindsight they can get blamed for not doing the other thing. Organizations often react counterproductively to the discovery of either form of trouble. When procedures are not followed but should have been, they may send more exhortations to follow procedures (or even more procedures) into the operation. When procedures were applied rigidly and followed by trouble, organizations may send signals to give operational people more discretion. But none of this resolves the fundamental double bind people are in, in fact, it may even tighten the bind. Either signal only shifts people's decision criterion a bit—faced with a difficult situation, the evidence in people's circumstances should determine whether they stick to the rules or adapt. But where the decision lies depends in large part on the signals the organization has been sending.

"Violation" of procedures, or "non-compliance" are obviously unproductive and judgmental labels—a form of saying "human error" all over again, without explaining anything. They simply rehearse that people should stick with the rules, and then everything will be safe. Ever heard of the "work to rule strike"? This is when people, instead of choosing to stop work altogether, mobilize industrial action by following all the rules for a change. What typically happens? The system comes to a gridlock. Follow all the rules by the book, and our systems no longer work.

Labels such as "violations" miss the complexity beneath the successful application, and adaptation, of procedures, and may lead to unproductive countermeasures. High reliability organizations do not try to constantly close the gap between procedures and practice by exhorting people to stick to the rules. Instead, they continually invest in their understanding of the reasons beneath the gap. This is where they try to learn—learn about ineffective guidance; learn about novel, adaptive strategies and where they do and do not work.

STRESS AND WORKLOAD

Stress has long been an important term, especially where people carry out dynamic, complex and safety-critical work. On a superficial reading of your mishap data, it may be easy to assert that people got stressed; that there was high workload and that things got out of hand because of it. But this does not mean or explain very much. Psychologists still debate whether stress is a feature of a situation, the mental result of a situation, or a physiological and psychological coping strategy that allows us to deal with a demanding situation. This complicates the use of stress in any causal statement, because what produced what?

Demand-resource mismatch

What you can do on the basis of your data is make an inventory of the demands in a situation, and the resources that people had available to cope with these demands. This is one way to handle your evidence. If you suspect that stress or high workload may have been an issue, look for examples of demands and resources in your situation (see Table 10.1).

In studies of stress and workload, some have reported that a mismatch between demands and resources may mean different things for different kinds of operators. In a marine patrol aircraft, for example, people in the back are concerned with dropping sonobuoys (to detect submarines) out of the aircraft. The more sonobuoys in a certain amount of time, the more workload, the more stress. People in the front of the aircraft were instead concerned with more strategic questions. For them, the number of things to do had little bearing on their experience of stress and workload. They would feel stressed, however, if their model of the situation did not match reality, or if it had fallen behind actual circumstances.

Curiously, a tiny mismatch between demands and resources may often lead to more feeling of stress than a really large mismatch. In other words, people will experience most stress when they have the idea that they can deal with the demands—get on top of it, so to say—by mustering just that extra bit of resources. This may happen, for example, after a change in plans. The new plan is not unmanageable (e.g. flying to a different runway), but requires more resources to be put in place than was originally counted on.

Tunneling and regression

One of the reported consequences of stress is tunneling—the tendency to see an increasingly narrow portion of one's operating environment. This is generally interpreted as a shortcoming; as something dysfunctional that marks less capable operators. Another consequence that has been noted is regression—the tendency to revert to earlier learned routines even if not entirely appropriate to the current situation.

But you can actually see both tunneling and regression as strategies in themselves; as a contributions from the human that are meant to deal with high demands (lots to pay attention to and keep track of) and limited resources (limited time too look around; limited mental workspace to integrate and deal with diverse and rapidly changing data). Tunneling (sometimes called "fixation", especially when people lock onto one explanation of the world around them) comes from the human strength to form a stable, robust idea of a shifting world with multiple threads that compete for attention and where evidence may be uncertain and incomplete. The threads that get people's attention may indeed be a limited set, and may consist of the threat (e.g. a system failure) rather than the process (e.g. flying the aircraft).

In highly dynamic and complex situations, it would seem that tunneling is an (involuntary) strategy that allows people to track and stay ahead of a limited number of threads out of a host of potential ones. Similarly, regression to earlier learned routines frees up mental resources: people do not have to match current perceptions with consciously finding out what to do each time anew.

Table 10.1: Finding a mismatch between problem demands and coping resources can help you make arguments about stress and workload more specific.

Problem demands:	Coping resources:
Ill-structured problems	Experience with similar problems
Highly dynamic circumstances: things changing quickly over time	Other people contributing to assessments of what is going on
Uncertainty about what is going on or about possible outcomes	Knowledge or training to deal with the circumstances
Interactions with other people that generate more investment than return (in terms of offloading)	Other people to off-load tasks or help solve problems
Organizational constraints and pressures	Organizational awareness of such pressures and constraints
Conflicts between goals	Guidance about goal priorities
High stakes associated with outcome	Knowledge there is an envelope of pathways to a safe outcome
Time pressure	Workload management skills

COORDINATION BREAKDOWNS

In most complex worlds, people do not carry out their work alone. Work, and the error detection and recovery in it, is inherently distributed over multiple people, likely in different roles.

- These people need to coordinate to get the work done
- Thus, problems in coordination may mark a sequence of events towards failure.

Crew Resource Management has become a popular label—not only in aviation, but also in medicine and other domains—that covers the coordinative processes between teammembers who pursue a common operational goal. So what does "the loss of effective CRM" mean? Here are some places to look for more specifics.

Differences between teammembers' goals

Complex operating environments invariably contain multiple goals that can all be active at the same time.

Take a simple flight from A to B: On-time arrivals, smooth rides through weather, slot allocation pressures, optimum fuel usage, availability of alternate airports, passenger convenience—these are all goals that can influence a single assessment or decision.

Given that people on the same operational team have different roles, not everyone on a team may feel equally affected by, or responsible for, some of these goals. This can lead to mismatches between what individuals see as their, or the team's, dominant pursuit at any one time.

Differences between teammembers' interpretation

Divergences can exist and grow in how people with different backgrounds and roles can interpret their circumstances. Different assessments can lead to the pursuit of different goals.

Gary Klein tells an interesting story of an airliner with three generators—one on each of its engines. One of the generators failed early in a flight. This is not particularly unsafe: two generators can provide the electrical power the aircraft needs. But then another engine began to lose oil, almost forcing a shut-down. After some discussion, the crew decided to let the ailing engine run idle, so that its generator could be called upon if necessary. When asked after landing how many generators had just been available, the co-pilot (who was flying the aircraft at the time) said "two". The captain said "one and a half", meaning one good engine and one idle. But the flight engineer said "one"--since getting the idle engine up and running where it powers the generator takes a moment.[6]

Knowledge that did not make it into the crew consciousness.

The story above also shows that certain knowledge can remain in a team's pre-conscious—that is, locked in the heads of individuals without being made public, or conscious. There may be many reasons why individuals do not contribute their understanding of the situation or their knowledge to the common ground, including overbearing commanders or shy subordinates.

But lack of coordination is often a matter of people assuming that others have a similar understanding of the situation; that others know what they themselves already know. Just like the flight engineer in the example above, who may have assumed that the two pilots knew how only one generator was available for at least a moment. Usually there are very good reasons for these assumptions, as they facilitate team coordination by not cluttering crew communication with redundant reminders and pointers. When you encounter differences between people's goals, between people's interpretations and when you find missing communications in the rubble, it is easy to look at them as failures or losses. Failures of teamwork, for example. Or failures of leadership, or loss of crew resource management. But look behind the failure. Silence by one crewmember may in actually represent good teamwork—which includes knowing when not to interrupt.

Features of the operating environment

Features of the operating environment can make the sharing of assessments and actions difficult. For example:

- Ergnomic problems such as high noise levels, low lighting or clumsy seating arrangements.
- More subtle features, especially related to computer technology, can impair coordination and cross-checking and the catching of errors.

Modern airliners are equipped with flight management systems (FMS's) that basically fly the entire aircraft. Pilots each have individual access to the FMS through a separate interface—their private little workspace. Here they can make significant changes to the flight plan without the other pilot necessarily seeing, knowing, or understanding. The pilot only needs to press "execute" and the computer will do what s/he has programmed.

Airlines have of course devised procedures that require pilots to cross-check each other's computer entries, but in reality there are many circumstances in which this is impractical or unnecessary. The real coordination problem is not pilots' failure to follow procedures. It is a feature of the design that makes coordination very difficult, yet safety-critical.[7]

LOOK IN HISTORY

The mishap you are investigating may seem the first of its kind in your organization, but chances are it is not. The potential of a mishap typically grows over time, and much can be learned from episodes of near-failure, or similar failures, that preceded the event you are examining now. In this section we will look at dress rehearsals and contrast cases.

Dress rehearsals

The period before a mishap may contain sequences of events that look like the one in the actual accident or incident, but without the same bad outcome. These could be called "dress rehearsals".

In January 1992, a highly automated aircraft crashed into a mountain close to Strasbourg airport in eastern France. Confusion between two automation modes that could each manage the aircraft's descent turned out to have been central in the crash. The pilots intended to make an automatic approach at a flight path angle of 3.3 degrees towards the runway. Due, however, to an internal connection between horizontal and vertical automation modes in the aircraft's computer systems, the aircraft was not in flight path angle mode, but had slipped into vertical speed mode. Pilots have to use the same knob in either mode, so dialing 3.3 resulted in a descent rate of 3300 feet per minute down—much steeper than 3.3 degrees.

During the years preceding this accident, various airlines had had similar sequences of events: aircraft flying in vertical speed mode instead of flight path angle mode. In these cases, go-arounds could be made. One airline had even developed some ad-hoc preventative training to avoid just this sort of event, even though it commented that pilots of the fleet were reluctant to admit there might be an ergonomic shortcoming in this cockpit.

Dress rehearsals tell you to look for more systemic contributors to the behavior in question. What are the commonalities? What is the trap that everybody seems to fall into? The contrast between dress rehearsal and actual mishap also shows what it takes to push a system over the edge, and what prevented a complete breakdown earlier.

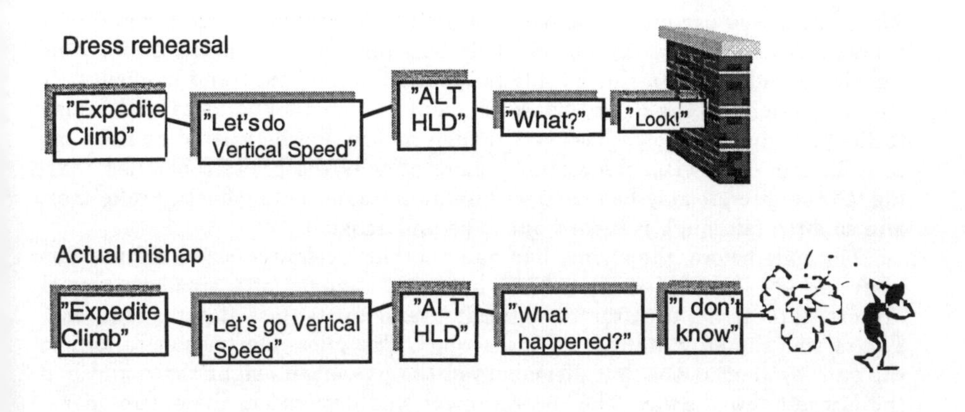

Figure 10.6: Dress rehearsal versus the actual mishap.

The Strasbourg crash happened at night, in snow. It is likely that the dress rehearsals took place in better conditions, where pilots had eye contact with the ground. Also, the airline going into Strasbourg had elected not to install Ground Proximity Warning Systems in its aircraft because of the high false alarm rate in the systems at that time, and the fact that it flew many short missions in mountainous terrain—exacerbating the false alarm problem. One dress rehearsals was kept from disaster by a Ground Proximity Warning.

That dress rehearsals can occur locally without subsequent investments in serious countermeasures also gives you a clue about an industry's perception of risk and danger, and reveals vulnerabilities in its way of sharing safety-critical information with other operators.

Contrast cases

Other mishaps, whether in the same organization or industry or not, can function as contrast cases. These are situations which are largely similar, but where people behaved slightly differently—making other assessments or decisions. This difference is a powerful clue to the reasons for behavior embedded in your situation.

An airliner was urgently requested by air traffic control to use a rapid exit taxiway from the runway on which it had just landed, because of traffic tightly behind it. The airliner could not make the final turn and momentarily slid completely off the hard surface. It reentered another taxiway and taxied to the gate under its own power. Although no procedures existed at the time to tell them otherwise, the airline wondered why the pilots continued taxiing, as the aircraft may have suffered unknown damage to wheels, brakelines, and so forth (although it turned out to be undamaged).

Not long before, the airline had had another incident where a similar aircraft had left the hard surface. This, however, occurred at a small provincial airport, late at night, after the aircraft's and pilots' last flight of the day. Theirs was the only aircraft on the airport. The pilots elected not to taxi to the gate by themselves, but disembarked the passengers right there and had the aircraft towed away. The control tower was involved in the entire operation.

This contrasted sharply with the other case, which happened at the airline's major hub. Many passengers had connecting flights, as did the pilots and their aircraft. It rained heavily, and the wind blew hard, making disembarkation on the field extremely undesirable. People in the control tower seemed not to have noticed the event. Moreover, for the time it would have taken to get busses and a tow truck out to the field, the aircraft would have blocked a major taxiway, all but choking the movements of aircraft landing behind it every two minutes.

In both contrast cases and dress rehearsals, it is important to focus on similarities, not differences. Operators apparently faced the same kinds of problems or dilemmas. Incidents mark vulnerabilities in how work is done, regardless of the outcome. Real progress on safety hinges on seeing the similarities between events, similarities that may highlight particular patterns toward breakdown (for example: the vertical speed mode instead of flight path angle) independent of the circumstances.

Notes

1 Material for this section comes from Woods, D. D., Johanssen, L. J., Cook, R. I., & Sarter, N. B. (1994). *Behind human error: Cognitive systems, computers and hindsight.* Dayton, OH: CSERIAC, and Dekker, S. W. A., & Hollnagel, E. (Eds.) (1999). *Coping with Computers in the Cockpit.* Aldershot, UK: Ashgate.

2 Cordesman, A. H., & Wagner, A. R. (1996). The lessons of modern war, Vol. 4: The Gulf War, Boulder, CO: Westview Press.

3 Vaughan, D. (1996). *The Challenger launch decision.* Chicago, IL: University of Chicago Press.

4 The quote on Murphy's law comes in part from Langewiesche, W. (1998). *Inside the sky.* New York: Pantheon.

5 Reason, J. (1997). Managing the risks of organizational accidents. Aldershot, UK: Ashgate Publishing Co.

6 Klein, G. (1998). *Sources of power: How people make decisions.* Cambridge, MA: MIT Press.

7 Dekker, S. W. A., & Hollnagel, E. (Eds.) (1999). *Coping with computers in the cockpit.* Aldershot, UK: Ashgate.

11. Writing Recommendations

Coming up with human factors recommendations can be one of the more difficult tasks in an investigation. Often only the shallowest of remedies seem to lie within reach. Tell people to watch out a little more carefully. Write another procedure to regiment their behavior. Or just get rid of the particular miscréants altogether. The limitations of such countermeasures are severe and deep, and well-documented:

- People will only watch out more carefully for so long, as the novelty and warning of the mishap wears off;
- A new procedure will at some point clash with operational demands or simply disappear in masses of other regulatory paperwork;
- Getting rid of the miscréants doesn't get rid of the problem they got themselves into. Others always seem to be waiting to follow in their footsteps.

A human error investigation should ultimately point to changes that will truly remove the error potential from a system—something that places a high premium on meaningful recommendations.

RECOMMENDATIONS AS PREDICTIONS

Coming up with meaningful recommendations may be easier if you think of them as predictions, or as a sort of experiment. Human error is systematically connected to features of the tasks and tools that people work with, and to features of the environment in which they carry out their work. Recommendations basically propose to change some of these features. Whether you want new procedures, new technologies, new training, new safety interlocks, new regulations, more managerial commitment—your recommendations essentially propose to re-tool or re-shape parts of the operational or organizational environment in the hope of altering the behavior that goes on within it.

In this sense your recommendations are a prediction, a hypothesis. You propose to modify something, and you implicitly predict it will have a certain effect on human behavior. The strength of your prediction, of course, hinges on the credibility of the connection you have shown earlier in your investigation: between the observed human errors and critical features of tasks, tools and environment. With this prediction in hand, you challenge those responsible for implementing your recommendations to go along in your experiment—to see if, over time, the proposed changes indeed have the desired effect on human performance.

High-end or low-end recommendations

So what about those changes? What kinds of changes can you propose that might have some effect on human performance? A basic choice open to you is how far up the causal chain you want your recommended changes to have an impact.

Typical of reactions to failure is that people start very low or downstream. Recommendations focus on those who committed the error, or on other operators like them. Recommendations low in the causal chain aim for example at retraining individuals who proved to be deficient, or at demoting them or getting rid of them in some other way. Other low-end recommendations may suggest to tighten procedures, presumably regimenting or boxing in the behavior of erratic and unreliable human beings.

Alternatively, recommendations can aim high—upstream in the causal chain—at structural decisions regarding resources, technologies and pressures that people in the workplace deal with. High-end recommendations could for example suggest to re-allocate resources to particular departments or operational activities.

This choice—upstream or downstream—is more or less yours as an investigator. And this choice directly influences:

- the ease with which your recommendation can be implemented;
- the effectiveness of your recommended change.

The ease of implementation and the effectiveness of an implemented recommendation generally work in opposite directions. In other words: the easier the recommendation can be sold and implemented, the less effective it will be (see figure 11.1).

Generally, recommendations for changes low on the causal chain are not very sweeping. They concentrate on a few individuals or a small subsection of an organization. These recommendations are satisfying for people who seek retribution for a mishap, or people who want to "set an example" by coming down on those who committed the errors.

But after implementation, the potential for the same kinds of error is left in the organization or operation. The error is almost guaranteed to repeat itself in some shape or form, through someone else who finds him-or herself in a similar situation. Low-end recommendations really deal with symptoms, not with causes. After their implementation, the system as a whole has not become much wiser or better.

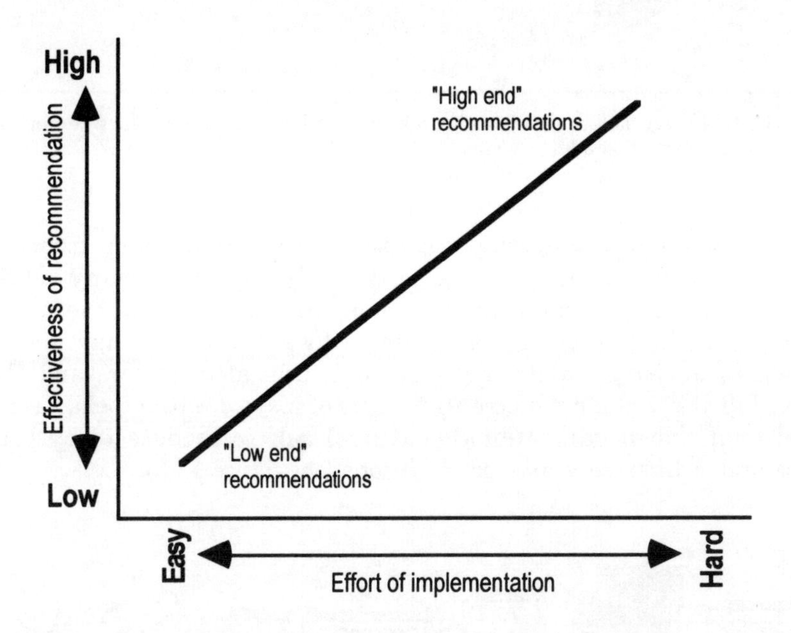

Figure 11.1: The trade-off between recommendations that will be easier to implement and recommendations that will actually have some lasting effect.

One reason for the illusion that low-end or other narrow recommendations will prevent recurrence is the idea that failure sequences always take a linear path: Take any step along the way out of the sequence, and the failure will no longer occur (see figure 11.2).

In complex, dynamic systems, however, this is hardly ever the case. The pathway towards failure is seldom linear or narrow or simple.

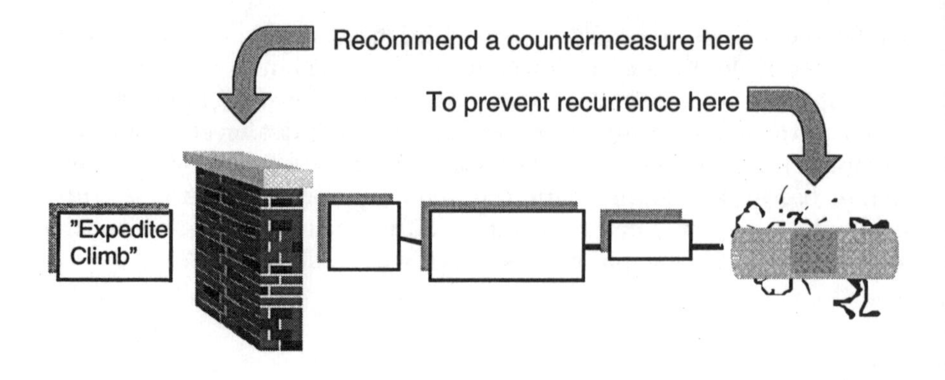

Figure 11.2: We may believe that blocking a known pathway to failure somewhere along the way will prevent all similar mishaps.

Mishaps have dense patterns of causes, with contributions from all corners and parts of the system, and typically depend on many subtle interactions. Putting one countermeasure in place somewhere along (what you thought was like) a linear pathway to failure may not be enough. In devising countermeasures it is crucial to understand the vulnerabilities through which entire parts of a system (the tools, tasks, operational and organizational features) can contribute to system failure under different guises or conditions (see figure 11.3).

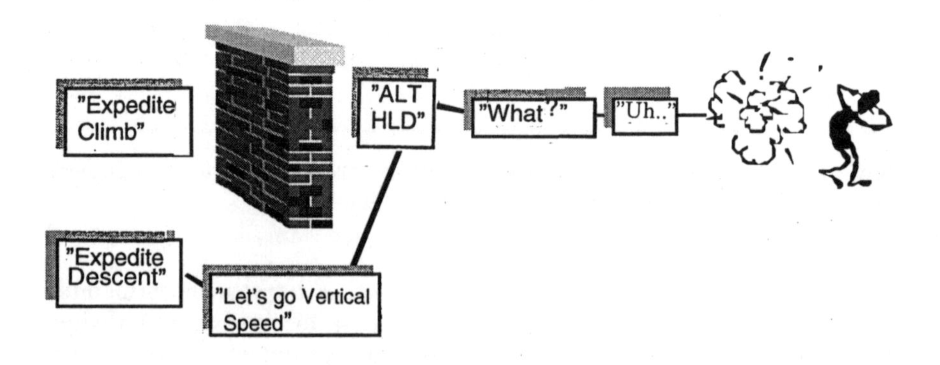

Figure 11.3: Without understanding and addressing the deeper and more subtle vulnerabilities that drive pathways towards failure, we leave opportunities for recurrence open.

Difficulties with high-end recommendations

The higher you aim in a causal chain, the more difficult it becomes to find acceptance for your recommendation. The proposed change will likely be substantial, structural or wholesale. It will almost certainly be more expensive. And it may concern those who are so far removed from any operational particulars that they can easily claim to bear no responsibility in causing this event or in helping to prevent the next one. Short of saying that it would be too expensive, organizations are good at finding reasons why structural recommendations do not need to be implemented, for example:

- "We already pay attention to that"
- "That's in the manual"
- "This is not our role"
- "We've got a procedure to cover that"
- "This recommendation has no relevance to the mishap"
- "People are selected and trained to deal with that"
- "This is not our problem."

It is easy to be put off as investigator before you even begin writing any recommendations. In fact, many recommendations that aim very high in the causal chain do not come out of first investigations, but out of re-opened inquiries, or ones re-submitted to higher authorities after compelling expressions of discontent with earlier conclusions.

One such case was the crash of a DC-10 airliner into Mount Erebus on Antarctica. The probable cause in the Aircraft Accident Report was the decision of the captain to continue the flight at low level toward an area of poor surface and horizon definition when the crew was not certain of their position. The kinds of recommendations that follow from such a probable cause statement are not difficult to imagine. Tighten procedures; exhort captains to be more careful next time around.

A subsequent Commission of Inquiry determined that the dominant cause was the mistake by airline officials who programmed the aircraft computers—a mistake directly attributable not so much to the persons who made it, but to the administrative airline procedures which made the mistake possible. The kinds of recommendations that follow from this conclusion would be different and aim more at the high end. Review the entire operation to Antarctica and the way in which it is prepared and managed. Institute double checking of computer programming. And so forth.[1]

The case for including or emphasizing high-end recommendations in a first investigation is strong. If anything, it is discouraging to have to investigate the same basic incident or accident twice. Structural changes are more likely to have an effect on the operation as a whole, by removing or foreclosing error traps that would otherwise remain present in the system.

Remember from chapter 4 that Judge Moshansky's investigation of the Air Ontario crash generated 191 recommendations. Most of these were high-end. They concerned for example:[2]

- Allocation of resources to safety versus production activities;
- Inadequate safety management by airline and authority alike;
- Management of organizational change;
- Deficiencies in operations and maintenance;
- Deficient management and introduction of new aircraft;
- Deficient lines of communication between management and personnel;
- Deficient scheduling (overcommitting this particular aircraft);
- Deficient monitoring and auditing;
- Deficient inspection and control and handling of information;
- Inadequate purchasing of spares;
- Low motivation and job instability following airline merger;
- Different corporate cultures;
- High employee turnover;
- Poor support to operational personnel;
- Inadequate policy making by airline and authority.

These are just some of the areas where recommendations were made. With a serious human error investigation, many of these kinds of conditions can probably be uncovered in any complex system. The ability to generate structural recommendations that aim high up in a causal chain is a reflection of the quality and depth of your understanding of human error.

SEARCHING THE EVIDENCE FOR COUNTERMEASURES

The kind and content of your recommendations depends, of course, on the kind and content of the mishap you are investigating. But to come up with high-end recommendations it may be useful to re-visit some of the organizational contributions to failure from chapter 9. For example:

- The re-allocation of resources that flow from the blunt end, and the alleviation of constraints that are imposed on operators' local decisions and trade-offs;
- Making goal conflicts explicit and turning them into topics for discussion among those involved;
- Re-invest in the defenses that turned out to be brittle or broken or non-existent;
- Make regulatory access more meaningful through a re-examination of the nature and depth of the relationship between regulator and operator.

Get help from the participants

If possible, it can be fruitful to build on the list above by talking to the participants themselves. These are some of the question that Gary Klein and his researchers ask participants when looking for countermeasures against recurrence of the mishap:

- What would have helped you to get the right picture of the situation?
- Would any specific training, experience, knowledge, procedures or cooperation with others have helped?
- If a key feature of the situation would have been different, what would you have done differently?
- Could clearer guidance from your company have helped you make a better trade-offs between conflicting goals?

Not only can answers to these questions identify countermeasures you perhaps had not yet thought of. They can also serve as a reality check. Would the countermeasures you think about proposing have any effect on the kind of situation you are trying to avoid? Asking the participants themselves, who after all have intimate knowledge of the situation you are investigating, may be a good idea.

PERSISTENT PROBLEMS WITH RECOMMENDATIONS

Recommendations are perhaps not the most satisfactory part of the investigator's job:

- It is not easy to formulate recommendations—especially with respect to human factors.
- It may be difficult to get recommendations implemented. Subtle wording may make the difference between rejection or acceptance, depending who sits on the committee reviewing the recommendations.
- It is difficult to find out whether implementation had any effect on the safety of the system.

Investigations are typically expensive. Organizations often allocate significant resources to probing an incident, regardless of whether they want to or have to because of regulations. The money spent on an investigation ups the stakes of the recommendations that come out of it. Apart from producing and circulating incident stories that may have a learning benefit (see chapter 12), recommendations are the only product to have some lasting effect in how an organization and its activities are designed or run. If all recommendations are rejected at the end, then the investigation has missed its goal.

One investigator described how the writing and inclusion of recommendations is heavily determined by who is going to be on the committee assessing the recommendations for implementation. Language may be adjusted or changed, some recommendations may be left out in order to increase the chances for others. This illustrates that the road from investigation to implementation of countermeasures is largely a political one.

It also reminds us of the disconnect between the quality of an investigation and its eventual impact on organizational practice. A really good investigation does not necessarily lead to the implementation of really good countermeasures. In fact, the opposite may be true if you look at figure 11.1. Really good investigations may reveal systemic shortcomings that necessitate fundamental interventions which are too expensive or sensitive to be accepted.

The focus on diagnosis, not change

Recommendations represent the less sexy, more arduous back-end of an investigation. One reason why they can fall by the wayside and not get implemented with any convincing effect is that they are fired off into an organization as limited, one-shot fixes. Many organizations—even those with mature safety departments and high investments in investigations—lack a coherent strategy on continuous improvement. Resources for quality control and operational safety are directed primarily at finding out what went wrong in the past, rather than assuring that it will go right in the future. The focus is on diagnosis, not change.

The emphasis on diagnosis can hamper progress on safety. Recommendations that are the result of careful diagnosis have not much hope of succeeding if nobody actively manageges a dedicated organizational improvement process. Similarly, feedback about the success of implemented recommendations will not generate itself. It needs to be actively sought out, looked for, compiled and sent back and assessed. Organizations seldom have mechanisms in place for generating feedback about implemented recommendations, and enjoy little support or understanding from senior management for the need for any such mechanisms. The issue, in the end, is one of sponsoring and maximizing organizational learning; organizational change, which is what chapter 12 is about.

Notes

1 See: Vette, G. (1983). *Impact Erebus.* Auckland, NZ: Hodder & Stoughton.
2 Moshansky, V. P. (1992). *Commission of inquiry into the Air Ontario accident at Dryden, Ontario* (Final report, vol. 1-4). Ottawa, ON: Minister of Supply and Services, Canada.

12. Learning from Failure

The point of any investigation is to learn from failure. Mishaps, in this regard, are a window of opportunity. The immediate aftermath of a mishap typically creates an atmosphere in which:

- Parts of an organization may welcome self-examination more than before;
- Traditional lines between management and operators, between regulators and operators, may be temporarily blurred in joint efforts to find out what went wrong and why;
- People and the systems they work in may be open to change—even if only for a short while;
- Resources may be available that are otherwise dedicated to production only, something that could make even the more difficult recommendations for change realistic.

Just doing the investigation, however, does not guarantee success in capitalizing on this window of opportunity. Learning from failure is about more than picking over the evidence of something gone wrong. Learning is about modifying an organization's basic assumptions and beliefs. It is about identifying, acknowledging and influencing the real sources of operational vulnerability. This can actually be done even before real failures occur, and the remainder of this chapter is about the opportunities and difficulties of organizational learning—before as well as after failures.

INVESTING IN A SAFETY CULTURE

Safety typically comes to the foreground only at certain moments—the frightening, surprising and generally expensive moments of mishaps. But it does not need to be that way. Organizations can invest in what is called a "safety culture", in order to learn about safety and its threats continually.

A SAFETY CULTURE IS ONE THAT ALLOWS THE BOSS TO HEAR BAD NEWS

The "easy" and "hard" problem of a safety culture

Creating a safety culture presents an organization with two problems: an easy one and a hard one. The easy problem (by no means easy, actually, but comparatively straightforward) is to make sure that bad news reaches the boss. Many organizations have instituted safety reporting systems that do exactly that: identifying and addressing problems before they can develop into incidents or accidents.

The hard problem is to decide what is bad news. Chapter 10, which discusses "complacency" as one label for human error, shows that an entire operation or organization can shift its idea of what is normative, and thus shift what counts as bad news. On-time performance can be normative, for example, even if it means that operators unknowingly borrow from safety to achieve it. In such cases, the hurried nature of a departure or arrival is not bad news that is worth reporting (or worth listening to, for that matter). It is the norm that everyone tries to adhere to since it satisfies other important organizational goals (customer service, financial gain) without obviously compromising safety.

Outside audits are one way to help an organization break out of the perception that its safety is uncompromised. In other words, neutral observers may better be able to spot the "bad news" among what are normal, everyday decisions and actions to people on the inside.

SIGNS OF NOT LEARNING FROM FAILURE:

Most organizations aim to learn from failures, either after they have happened or before they are about to happen. The path to learning from failure is generally paved with intentions to embrace the new view of human error; to see human error as a symptom of deeper, systemic trouble. But many obstacles get in the way, frustrating attempts to learn—either after a serious failure or on the way towards one. Here are some signs of people not learning—all are ways in which organizations try to limit the need for fundamental change.

"To err is human"

Although it is a forgiving stance to take, organizations that suggest that "to err is simply human" may normalize error to the point where it is no longer interpreted as a sign of deeper trouble.

"There is one place where doctors can talk candidly about their mistakes. It is called the Morbidity and Mortality Conference, or more simply, M. & M. Surgeons, in particular, take M. & M. seriously. Here they can gather behind closed doors to review the mistakes, complications and deaths that occurred on their watch, determine responsibility, and figure out what to do differently next time."

A sophisticated instrument for trying to learn from failure, M. & M.'s assume that every doctor can make errors, yet that no doctor should—avoiding errors is largely a matter of will. This can truncate the search for deeper, error-producing conditions. In fact, "the M & M takes none of this into account. For that reason, many experts see it as a rather shabby approach to analyzing error and improving performance in medicine. It is isn't enough to ask what a clinician could or should have done differently so that he and others may learn for next time. The doctor is often only the final actor in a chain of events that set him or her up to fail. Error experts, therefore, believe that it's the process, not the individuals in it, which requires closer examination and correction."[1]

"Setting examples"

Organizations that believe they have to "set an example" by punishing or reprimanding individual operators are not learning from failure. The illusion is there, of course: if error carries repercussions for individuals, then others will learn to be more careful too.

The problem is that instead of making people avoid errors, an organization will make people avoid the reporting of errors, or the reporting of conditions that may produce such errors.

In one organization it is not unusual for new operators to violate operating procedures as a sort of "initiation rite" when they get qualified for work on a new machine. By this they show veteran operators that they can handle the new machine just as well. To be sure, not all new operators take part, but many do. In fact, it is difficult to be sure how many take part. Occasionally, news of the violations reaches management, however. They respond by punishing the individual violators (typically demoting them), thus "setting examples".

The problem is that instead of mitigating the risky initiation practice, these organizational responses entrench it. The pressure on new operators is now not only to violate rules, but to make sure that they aren't caught doing it—making the initiation rite even more of a thrill for everyone. The message to operators is: don't get caught violating the rules. And if you do get caught, you deserve to be punished—not because you violate the rules, but because you were dumb enough to get caught.

A proposal was launched to make a few operators—who got caught violating rules even more than usual—into teachers for new operators. These teachers would be able to tell from their own experience about the pressures and risks of the practice and getting qualified. Management, however, voted down the proposal because all operators expected punishment of the perpetrators. "Promoting" them to teachers was thought to send entirely the wrong message: it would show that management condoned the practice.

Compartmentalization

One way to deal with information that threatens basic beliefs and assumptions about the safety of the system is to compartmentalize it; to contain it.

In the organization described above, the "initiation rite" takes place when new operators are qualifying for working on a new machine. So, nominally, it happens under the auspices of the training department. When other departments hear about the practice, all they do is turn their heads and declare that it is a "training problem". A problem, in other words, of which they have no part and from which they have nothing to learn.

The problem is that compartmentalization limits the reach of safety information. The assumption beneath compartmentalization is that the need to change—if there is a need at all—is an isolated one: it is someone else's problem. There is no larger lesson to be learned (about culture, for example) through which the entire organization may see the need for change. In the example above, were not all operators—also all operators outside the training department—once new operators, and thus maybe exposed to or affected by the pressures that the initiation rite represents?

What seems to characterize high reliability organizations (ones that invest heavily in learning from failure) more than anything is the ability to identify commonalities across incidents. Instead of departments distancing themselves from problems that occur at other times or places and focusing on the differences and unique features (real or imagined), they seek similarities that contain lessons for all to learn.

Blaming someone else: the regulator for example

Most safety-critical industries are regulated in some way. With the specific data of an accident in hand, it is always easy to find gaps where the regulator "failed" in its monitoring role. This is not a very meaningful finding, however. Identifying regulatory oversights in hindsight does not explain the reasons for those—what now look like—obvious omissions. Local workload, the need to keep up with ever-changing technologies and working practices and the fact that the narrow technical expertise of many inspectors can hardly foresee the kinds of complex, interactive sequences that produce real accidents, all conspire against a regulator's ability to exercise its role. If you feel you have to address the regulator in your investigation, do not look for where they went wrong. As with investigating the assessments and actions of operators, find out how the regulator's trade-offs, perceptions and judgments made local sense at the time; why what they were doing or looking at was the right thing given their goals, resources, and understanding of the situation.

Another complaint often leveled against regulators is that they collude with those they are supposed to regulate, but this is largely a red herring (and, interestingly, almost universally disagreed with by those who are regulated. Independent of claims to collusion, they often see regulators as behind the times, intrusive and threatening). To get the information they need, regulators are to a large extent dependent on the organizations they regulate, and likely even on personal relationships with people in those organizations. The choice, really, is between creating an adversarial atmosphere in which it will be difficult to get access to safety-related information, or one in which a joint investment in safety is seen as in everybody's best interest.

As soon as you "pass the buck" of having to learn from a mistake to someone else—another person, another department, another company (e.g. suppliers), another kind of organization (e.g. regulator versus operator), you are probably shortchanging yourself with respect to the lessons that are to be learned. Failure is not someone else's problem.

OBSTACLES TO LEARNING

From the foregoing section, you may be able to recognize the signs of organizations not learning from failure. But why don't they learn? Apart from the reasons already mentioned in chapter 1 (resource constraints, reactions to failure, hindsight bias, limited human factors knowledge), there are more institutionalized obstacles to learning from failure.

Management were operators themselves

What characterizes many safety-critical organizations is that senior managers were often operators themselves—or still are (part-time). For example, in hospitals, physicians run departments, in airlines pilots do. On the one hand this provides an opportunity. Managers can identify with operators in terms of the pressures and dilemmas that exist in their jobs, thus making it easier for them to get access to the underlying sources of error.

But it can backfire too. The fact that managers were once operators themselves may rob them of credibility when it comes to proposing fundamental changes that affect everyone.

The organization in the examples above is one where senior management is made up of operators or ex-operators. What if management would want to reduce the risks associated with the initiation practice, or eliminate it altogether? They were once new operators themselves and very likely did the same thing when getting qualified. It is difficult for them to attain credibility in any proposal to curb the practice.

Over-zealous safety management

Sometimes the formal process of investigating mishaps and coming up with recommendations for change may itself stand in the way of learning from failure. In the aftermath of failure, the pressure to come up with findings and recommendations quickly can be enormous—depending on the visibility of the industry. An intense concern for safety (or showing such concern) can translate into pressure to reach closure quickly, something that can lead to a superficial study of the mishap and its deeper sources.

Also, concern for safety in a company or across an industry can promote the creation of safety departments and safety specialists. There have been cases where safety professionals have become divorced from daily operations to an extent where they only have a highly idealized view of the actual work processes and are no longer able to identify with the point of view of people who actually do the safety critical work every day.

Litigation

It is becoming increasingly normal—and very worrying to large segments of the safety community—that operators involved in mishaps get sued or charged with (criminal) offenses.

Valujet flight 592 crashed after take-off from Miami airport because oxygen generators in its cargo hold caught fire. The generators had been loaded onto the airplane by employees of a maintenance contractor, who were subsequently prosecuted. The editor of Aviation Week and Space Technology "strongly believed the failure of SabreTech employees to put caps on oxygen generators constituted willful negligence that led to the killing of 110 passengers and crew. Prosecutors were right to bring charges. There has to be some fear that not doing one's job correctly could lead to prosecution."[2]

But prosecution of individuals misses the point. It shortcuts the need to learn fundamental lessons, if it acknowledges that fundamental lessons are there to be learned in the first place. In the SabreTech case, the lowly maintenance employees inhabited a world of boss-men and sudden firings, stumbled through an operation that did not supply safety caps for expired oxygen generators and in which the airline was as inexperienced and under as much financial pressure as people in the maintenance organization supporting it. It was also a world of language difficulties—not only because many were Spanish speakers in an environment of English engineering language:

"Here is what really happened. Nearly 600 people logged work time against the three Valujet airplanes in SabreTech's Miami hangar; of them 72 workers logged 910 hours across several weeks against the job of replacing the "expired" oxygen generators—those at the end of their approved lives. According to the supplied Valujet work card 0069, the second step of the seven-step process was: 'If the generator has not been expended install shipping cap on the firing pin.'

This required a gang of hard-pressed mechanics to draw a distinction between canisters that were 'expired', meaning the ones they were removing, and canisters that were not 'expended', meaning the same ones, loaded and ready to fire, on which they were now expected to put nonexistent caps. Also involved were canisters which were expired and expended, and others which were not expired but were expended. And then, of course, there was the simpler thing—a set of new replacement canisters, which were both unexpended and unexpired."[3]

These were conditions that existed long before the Valujet accident, and that exist in many places today. Fear of prosecution stifles the flow of information about such conditions. And information is the prime asset that makes a safety culture work. A flow of information earlier could in fact have told the bad news. It could have revealed these features of people's tasks and tools; these long-standing vulnerabilities that form the stuff that accidents are made of. It would have shown how human error is inextricably connected to how the work is done, with what resources, and under what circumstances and pressures.

Notes

1 Gawande, A. (1999). When doctors make mistakes. *The New Yorker*, February 1, pages 40-55.
2 North, D. M. (2000). Let judicial system run its course in crash cases. *Aviation Week and Space Technology, May 15*, page 66.
3 Langewiesche, W. (1998). *Inside the sky*. New York: Random House, page 228.

13. Rules for in the Rubble

What from this Field Guide should you take into the field with you? Faced with rubble of human error, what core rules of engagement should you remember? This chapter is a compilation of the most important lessons from the Field Guide. It gives you a brief overview of the do's and don'ts of a proper human error analysis.

The outcome of a sequence of events is most likely the reason why you are conducting an investigation in the first place. If there is no adverse outcome (or no likelihood of one)—there is no investigation. Being there because of the outcome has a problem to it: you *know* the outcome, or suspect the likely outcome. The problem is that your knowledge of outcome clouds your ability to evaluate human performance. Therefore, Rule of engagement 1 says:

1 You cannot use the outcome of a sequence of events to assess the quality of the decisions and actions that led up to it.

Once you know the outcome, you cannot pretend that you don't know it, of course. But you can remind yourself of the systematic effects of the hindsight bias and try to keep them out of your investigation:

2 Don't mix elements from your own reality now into the reality that surrounded people at the time. Resituate performance in the circumstances that brought it forth and leave it there.
3 Don't present the people you investigate with a shopping bag full of epiphanies ("look at all this! It should have been so clear!"), because that is not the way evidence about the unfolding situation reached people at the time.
4 Recognize that consistencies, certainties and clarities are products of your hindsight—not data available to people inside the situation. That situation was most likely marked by ambiguity, uncertainty and various pressures.
5 To understand and evaluate human performance, you must understand how the situation unfolded around people at the time, and take on the view from inside that situation. From there, try to understand how actions and assessments could have made sense.
6 Remember that the point of a human error investigation is to

understand why people did what they did, not to judge them for what they did not do.

Despite these rules of engagement, errors in investigating human error are easily made. All of them actually stem from the hindsight bias in one way or another. Try not to make these errors:

- **The counterfactual reasoning error**. You will say what people could or should have done to avoid the mishap. ("If only they..."). Saying what people did not do but could have done does not explain why they did what they did;
- **The data availability-observability error**. You will highlight the data that was available in the world surrounding people and wonder how they could have possibly missed it. Pointing out the data that would or could have revealed the true nature of the situation does not explain people's interpretation of the situation at the time. For that you need to understand which data was observed or used and how and why.
- **The micro-matching error**. You will try to match fragments of people's performance with all kinds of rules and procedures that you excavate from written documentation afterward. And of course, you will find gaps where people did not follow procedures. This mismatch, however, does not at all explain why they did what they did. And, for that matter, it is probably not even unique to the sequence of events you are investigating.
- **The cherry-picking error**. You identify an over-arching condition in hindsight ("they were in a hurry"), based on the outcome, and trace back through the sequence of events to prove yourself right. This is a clear violation of rule of engagement 2: leave performance in the context that brought it forth. Don't lift disconnected fragments out to prove a point you can really only make in hindsight.

At all times, remember the local rationality principle. People are not unlimited cognitive processors (in fact, there is not a single unlimited cognitive processor in the entire world, neither machine nor human). They cannot know everything at all times. People are able to deal with a limited number of cues, indications, evidence, knowledge, goals. What people do makes sense from their point of view, their knowledge, their objectives and limited resources.

The bottom line is this: people's actions and assessments *have* to make sense when viewed from their position inside the situation. If, despite your best efforts, you cannot have people's actions and assessments make sense, then human factors is probably not the field you want to be looking at for guidance on how to explain their behavior. You may need to go to psychiatry or clinical psychology if you really cannot make sense out of certain decisions or actions. Those fields could help you explain behavior on the basis of suicidal tendencies, for example. Or you may have to go to criminology if you find evidence of deliberate sabotage. As long as human performance can be made to make sense, using human factors concepts (and most performance in complex dynamic worlds can), you are safe within this field and within this Field Guide. Push on performance until it makes sense. Because it probably will.

Here is a brief reprise of the steps explained in chapter 9—steps that can help you make sense of the behavior in your mishap:

Step 1. Lay out the sequence of events based on the data you have gathered. You can use language of the domain (a context-specific language) in which the mishap occured to structure the events, and use time (and space) as principles along which to organize them.

Step 2. Divide the sequence of events into episodes that you can study separately for now. Each of these episodes may fit a different human factors explanation, but you may also find that you have to re-adjust the boundaries of your episodes later on.

Step 3. Reconstruct critical features of the situation around each of these events. What did the world look like, what was the process doing at the time? What data were available to people?

Step 4. Identify what people were doing or trying to accomplish at each episode. Reconstruct which data were actually observable. See what goals people were pursuing, what knowledge they used, and where, as a consequence, their attention was focused. Be relentless: press on their behavior until it makes sense to you.

Step 5. Link the details of your sequence of events to human factors concepts. In other words, build an account of the sequence of events that runs parallel to the account in step 1, but is instead cast in concept-dependent, or human factors terms. This will help you synthesize across mishaps and understand broader patterns of failure.

So how can you know that you have got it right? You can't, really. Each story we build to explain past performance is always tentative. You should be suspicious of anyone who claims he or she knows "exactly" what happened, and you should be suspicious if you yourself have that feeling about your investigation. New data may prove you wrong; new interpretations may be better than what you came up with. This is why it is so important to leave a trace; to reconstruct the situation in which people found themselves and use *that* as your starting point for modeling the psychological; and to avoid folk modeling. If you want to give other people a chance to review and re-interpret what you have done, you have to leave them with as much as possible.

Finally, remember to see each "human error"—under whatever traditional or fashionable label (bad decision, loss of situation awareness, violation, managerial deficiency, regulator failure)—as the *beginning* of your investigation, or the beginning of a new part, not as the conclusion. Indeed, replacing operator error with designer or organizational or regulator error does not lead you anywhere, other than to the beginning of yet another investigation. You have to resist pointing to other people as the source of failure, whoever they are or wherever they are. When you encounter other "human errors", elsewhere, you are back at step 1. Error, by any other name, or by any other human, is the start of your probe, not its conclusion.

SINNERS VERSUS SAINTS; EXCUSING VERSUS EXPLAINING

The Field Guide may be, to some, a conspiracy to get culprits off the hook. The steps it contains may amount, to some, to one big exculpatory exercise—explaining away error; making it less bad; less egregious; less sinful; normalizing it.

The reaction is understandable. Some people always need to bear the brunt of a system's failure. These are often people at the blunt end of organizations. Managers, supervisors, boardmembers: *they* have to explain the failure to customers, patients, passengers, stockowners, lawyers. The Field Guide may look like a ploy to excuse defective operators; to let them plod on with only minor warnings or no repercussions at all, as uncorrected unreliable elements in an otherwise safe system. The Field Guide, however, aims to help explain the riddle of puzzling human performance—not explain it away. There is a difference between explaining and excusing human performance. The former is what the Field Guide helps you do. The latter is not within its purview, but is something that hinges on the norms and laws that govern practice and reactions to failure within your organization or domain. In fact, the whole exculpatory argument invokes an old debate that really leads us nowhere. The question in this debate is:

- Is human failure part of our nature? Are we born sinners in otherwise safe systems?
- Or is it a result of nurture? Are we saints at the mercy of deficient organizations, not quite holding our last stand against systems full of latent failures?

People are neither just sinners nor simply saints—at best they are both. The debate, and the basic distinction in it, oversimplifies the challenges before us if we really want to understand the complex dynamics that accompany and produce human error. Safety (and failures) are emerging features; they are by-products of entire systems and how these function in a changing, uncertain and resource-constrained world. Neither safety nor failure is the prerogative of select groups or individuals inside of these systems.

This is why the pursuit of culprits—organizational or individual—does not work. A system cannot learn from failure and punish supposedly responsible individuals or groups at the same time. Although you may think that they go hand in hand, the two are really mutually exclusive:

- Punishing is about keeping your beliefs in a basically safe system intact (the only threat comes from the people you are now punishing, or from those who could otherwise take an example from them). Learning is about changing these beliefs, and changing the system accordingly.
- Punishing is about seeing the culprits as unique parts of the failure, as in: it would not have happened if it were not for them. Learning is about seeing the failure as a part of the system.
- Punishing is about teaching your people not to get caught the next time. Learning is about countermeasures that remove error-producing conditions so there won't be a next time.
- Punishing is about stifling the flow of safety-related information (because people do not want to get caught). Learning is about increasing that flow.
- Punishing is about closure, about moving beyond the terrible event. Learning is about continuity, about the continuous improvement that comes from firmly integrating the terrible event in what the system knows about itself.

Acknowledgements

The Field Guide to Human Error Investigations was born through participation in various incident and accident investigations. I want to thank those who alerted me to the need for this book and who inspired me to write it, in particular Air Safety Investigator Maurice Peters and Captain Örjan Goteman. It was written on a grant from the Swedish Flight Safety Directorate and Arne Axelsson, its director. Kip Smith and Captain Robert van Gelder and his colleagues were invaluable for their comments and suggestions during the writing of earlier drafts.

I am indebted to the following people for "the new view" on human error: David Woods, Erik Hollnagel, Edwin Hutchins, James Reason, Charles Billings, John Flach, Gary Klein, Judith Orasanu, Diane Vaughan, Gene Rochlin and Nick McDonald. The ideas in The Field Guide are inspired by them and their ideas, although any misrepresentations or biases in this book are of course my responsibility.

S.D.
Linköping, Sweden
Summer 2001

Subject Index